Dictionary of
Colloid and
Surface Science

Dictionary of Colloid and Surface Science

PAUL BECHER
Paul Becher Associates Ltd.
Wilmington, Delaware

CRC Press
Taylor & Francis Group
Boca Raton London New York

CRC Press is an imprint of the
Taylor & Francis Group, an **informa** business

This book was composed by the author, using a Dell 310 Computer and WordStar 5.0 and 5.5. It was printed on a Hewlett-Packard Laserjet Series II. The typeface is Times Roman and Math Times Roman, 10-point-on-12.

First published 1990 by Marcel Dekker, Inc.

Published 2009 by CRC Press
Taylor & Francis Group
6000 Broken Sound Parkway NW, Suite 300
Boca Raton, FL 33487-2742

ISBN 13: 978-0-367-45087-8 (pbk)
ISBN 13: 978-0-8247-8326-6 (hbk)

**Visit the Taylor & Francis Web site at
http://www.taylorandfrancis.com**

**and the CRC Press Web site at
http://www.crcpress.com**

Library of Congress Cataloging-in-Publication Data

Becher, Paul.
 Dictionary of colloid and surface science / Paul Becher.
  p. cm.
 Includes bibliographical references.
 ISBN 0-8247-8326-3 (alk. paper)
  1. Colloids--Dictionaries. 2. Surface chemistry--Dictionaries.
 3. Surfaces (Technology)--Dictionaries. 4. Surfaces (Physics)-
 -Dictionaries. I. Title.
 QD549.B413 1990
 541.3'45'03--dc20   89-27399
                                                 CIP

Preface

*But how little reason have we to
boast of our knowledge, when we
only gaze at the surfaces of things?*
—Samuel Johnson

As the reader might gather from the above rubric, I have long
been an admirer of Samuel Johnson: the canard of his "sesqui-
pedalian ponderosity" is dissipated by reading his works. I have
admired Johnson for his simple insistence on using the precisely
right word in precisely the right place. This insistence is, after
all, the mark of the lexicographer.

Thus, having been offered the opportunity of myself becoming a lexicographer, I first considered Johnson's own definition
of this position ("a harmless drudge"), and then my desire to
emulate this great man. There was, after all, no contest. I am
willing to be a harmless drudge in the pursuit of my profession.
Hence, this dictionary.

By turning a few pages, however, the reader will recognize
that this is not written as a standard dictionary. Rather, it is a
mini-encyclopedia. In many cases, the definition includes
some historical matter, or philosophical sidelight on the impli-
cations of the definition. I have also included brief biographical notes, sometimes in the body of the definition, but also
(when the scientist has been a major contributor to colloid and
surface science) I have added a biographical sketch under the

scientist's name. Cross-references are printed in **boldface**.

Such a work would not have been possible without help. First of all, I must thank Prof. M. Clausse, who first suggested the need for such a dictionary. I would also express my thanks to those who suggested and supplied definitions: D. Z. Becher, H. T. Davis, G. L. Gaines, Jr., R. G. Good, R. G. Laughlin, S. Ross, C. J. van Oss. Valuable biographical data was supplied by S. Ross and J. Th. G. Overbeek. Prof. A. Adamson has kindly supplied the key-word list used by the journal *Langmuir*. Other sources are listed in Appendix G - Bibliography.

I would finally like to express my thanks to Stuart Berg Flexner (who was the first real lexicographer I ever met!), for giving me some appreciation of what lexicography was all about, and for some encouraging words at the outset of this project.

And, of course, to Jane, who puts up with this kind of thing.

It only remains to mention the obvious - errors and omissions are entirely attributable to the author, who, however, would be grateful to have them pointed out.

Paul Becher

Contents

Dictionary of Colloid and Surface Science

A

absorbance *n* In optics, the term in the **Beer-Lambert law** characteristic of the substance whose light absorption is being measured.

absorbate *n* The species absorbed. See **absorption**.

absorbent *n* The species in which the **absorbate** is absorbed. See **absorption**.

absorption *n* Sorption in which the sorbed species (**absorbate**) penetrates the surface, and enters into the *bulk* of the **absorb-ent** phase. See **adsorption**.

acronym *n* A word, made up of the initial letters of a phrase, and usually written in capital letters, to be used as a short-hand description. Usually designed to be pronounceable, e.g., WAC for *W*omen's *A*rmy *C*orps, but not always, e.g., XRD for *X*-*R*ay *D*iffraction. In this Dictionary, definitions of acronyms are usually to be found under the acronym.

activation energy The minimum amount of energy which must be supplied to a system in order that a process may take place. Most commonly, in respect to chemical reactions; but also with reference to others, e.g., energy of **flocculation**. See **potential barrier**.

active site On a solid, locations particularly susceptible to **adsorption**. See **adsorption sites**.

activity, mean ionic See **mean ionic activity**.

Adam, Neil Kensington (1891-1973) British scientist known for his pioneering monolayer studies; author of the magisterial *The Physics and Chemistry of Interfaces*.

Adams, John Couch (1819-1892) British mathematician. See also **Bashforth, Francis**.

adatom *n* An atom **adsorbed** to a surface, e.g., the face of a crystal.

adhesion *n* The bonding of one body to another; the condition under which mechanical force or work may be transferred from one solid body to another, in tension or in shear, without interfacial slip or inelastic displacement of one body with respect to the other. See **adhesive**.

adhesion, energy of The intermolecular energy of attraction between the separate molecules making up a homogeneous liquid or solid. The energy consists of contributions from **Lifshitz-van der Waals** components in all cases; it may also have a **polar** component (in polar liquids or solids); and a metal component in metals. For a liquid/liquid or liquid/solid **interface**, it may be estimated as equal to the sum of the individual **surface tensions** less the **interfacial tension**, i.e.,

$$W_A = \gamma_1 + \gamma_2 - \gamma_{12}$$

This equation is ascribed to **Dupré**.

adhesive *n* A material which enables bodies to adhere; a glue. —*adj* pertaining to **adhesion**, as in *adhesive composition*.

adsorbate *n* The species which is adsorbed. When the adsorbate is a solute, it may be, for example, a **surface-active agent** or **surfactant**.

adsorbent *n* The solid on whose surface **adsorption** occurs.

adsorption *n* Adsorption is **sorption** in which the sorbed species accumulates in an **interfacial layer**. Note that it is possible for adsorption and **absorption** to occur simultaneously. See **adsorption isotherm**.

adsorption, heat of The heat evolved, especially in gas/solid adsorption. Depending on the conditions of measurement, the heat determined may be an *integral*, *differential*, or *isosteric* quantity.

adsorption hysteresis The phenomenon in which the **adsorption isotherm** is not coincident with the **desorption** isotherm; commonly observed in porous solids.

adsorption isostere See **isostere**.

adsorption isotherm An equation, usually theoretical, but sometimes more-or-less empirical, relating the amount of material adsorbed on a solid to the concentration of the **adsorbate**. See **BET; Freundlich isotherm; Gibbs isotherm; Langmuir isotherm; Stern isotherm**.

adsorption, negative See **negative adsorption**.

adsorption potential See **Polanyi adsorption isotherm**.

adsorption sites The positions on a solid to which an **adsorbate** may be **adsorbed**. The *number* of adsorption sites governs the number of molecules which constitute a **monolayer**. See **active site**.

adsorptive *n* See **adsorbate**. —*adj*, as in *adsorptive properties*.

advancing contact angle The **contact angle** observed as the liquid meniscus moves as a result of addition of liquid to the solid surface, or the larger contact angle observed when the solid surface is tilted. See **contact angle hysteresis; receding**

contact angle.

aerosol *n* A colloidal dispersion in a gas. The dispersed particles may be either solid or liquid; hence one may speak of *aerosols of solid (or liquid) particles.* The usage *solid (liquid) aerosol* is to be discouraged. Other descriptive terms include *fog* and *smoke.*

AES *acronym* *A*uger *e*lectron *s*pectroscopy. When a surface is scanned with an electron beam at a grazing angle, an excited atom may emit an inner electron; outer electrons may fall into the vacated state, leading to the emission of x-rays or of another electron (Auger effect). Since the energies of the electrons thus emitted are characteristic of the elements encountered by the beam, AES may be used for the chemical analysis of the surface. Frequently used in conjunction with **LEED** analysis. Named for Pierre V. Auger (b. 1899), French physicist. See also **SEM.**

agglomerate *n* A collection of primary particles or aggregates joined at their *edges* or *corners*, in such a way that the **specific surface area** is not markedly different from the sum of the areas of the constituent particles. —*v* the act of agglomeration. See **aggregate; floc.**

aggregate *n* A collection of primary particles joined at their *faces*, with a **specific surface area** significantly less than the sum of the areas of the constituent particles. Aggregates are distinguished from **agglomerates** by the increased difficulty in separation. —*v* the act of aggregation. See **floc; perikinetic; orthokinetic.**

aggregation number The number of **surfactant** molecules making up a **micelle**.

Amontons' Law The basic law of **friction**, first stated by Guillaume Amontons (1663-1705), French engineer, to the effect that the coefficient of friction μ is independent of the area of contact, i.e.,

$$\mu = F/W$$

where F is the frictional force, and W is the load or force normal to the direction of motion. See **lubrication**.

amphipathic *adj* Describing a molecule combining **hydrophilic** and **lipophilic** properties. Coined by G. S. Hartley.

amphoteric *adj* Capable of acting as an acid or a base; applied to a **surfactant**, capable of behaving as an **anionic** or **cationic**, depending (usually) on the pH. —*n* Briefly, an **amphoteric surfactant**.

anchor *n* In **steric stabilization**, that portion of the polymeric stabilizer which *anchors* the stabilizing molecule to the surface.

Andreasen pipette Device to measure **particle size distribution**, consisting of a graduated cylinder with an arrangement for withdrawing samples from the bottom at various times. Analysis of the samples and application of **Stokes' law** permits determination of the distribution.

anionic; anionic surfactant *n* An ionic **surface-active agent**

(**surfactant**) in which the surface-active moiety is the anion, e.g., sodium stearate, $C_{17}H_{35}COO^-$ Na^+. See **cationic; nonionic**.

anneal *v* To heat a substance to a temperature just below the bulk melting point.

anodic branch That portion of the **electrocapillarity** curve to the left of the maximum, i.e., the portion of the curve in which the mercury electrode is positively charged. See **cathodic branch**.

anomalous water See **water, anomalous**.

antifoam *n* A substance used to destabilize, or inhibit the formation of, **foam**. Commercial antifoams are usually some form of poly(dimethylsiloxane) polymer, but short-chain alcohols are frequently effective. The inhibition or destabilization of unwanted foam can be a commercial problem of large magnitude.

Antonoff Rule An empirical equation for calculating the **interfacial tension** between two liquids

$$\gamma_{12} = |\gamma_1 - \gamma_2|$$

where γ_{12} is the interfacial tension, and γ_1 and γ_2 the respective **surface tensions**. There are a number of limitations to this rule, usually not stated: the phases must be mutually saturated, and the **spreading coefficient** must be zero. Since the latter condition is rarely met, the rule provides only the roughest estimate. See **Good-Girifalco equation; Fowkes equation.**

apolar *adj* Characteristic of a substance or a surface devoid of any **polar** or metal property. Pertaining to low-energy substances or surfaces whose cohesive or adhesive interactions are governed solely by **Lifshitz-van der Waals** interactions.

aprotic *adj* Not containing a dissociable proton, e.g., benzene.

Argand diagram A Cartesian plot of complex numbers, the real part being plotted on the horizontal axis, and the imaginary part along the vertical axis. For Jean-Robert Argand (1768-1822), Swiss mathematician.

arithmetic mean See Appendix B.

association colloid A **surface-active agent** capable of forming **micelles**, i.e., of *associating*.

Auger electron spectroscopy See AES.

autophobic *adj* To describe films formed on a solid by deposition from a monolayer, having the property of *not* being **wetted** by the liquid from which they were deposited. The effect presumably arises from the fact that the outward-facing tails of the deposited molecules terminate in methyl groups, whose low **critical surface tension** effectively precludes wetting. See **Langmuir-Blodgett film.**

average *n* A quantity representative of the typical value of a set, e.g., a series of measurements. See Appendix B.

Avogadro's Law The principle that equal volumes of gases at the same pressure and temperature contain the same number of molecules. See **Avogadro number**.

Avogadro number The number of molecules N in a mole, equal to 6.022×10^{23}. For Count Amadeo Avogadro (1776-1856), Italian chemist and physicist. Also (esp. in German literature) **Loschmidt number**.

axial ratio The ratio of the two principal axes of a nonspherical particle, e.g., an **ellipsoid** (the sphere is, of course, a special case, with an axial ratio of one).

B

Bancroft, Wilder Dwight (1867-1953). American chemist. Founder of the *Journal of Physical Chemistry* (for some reason, the American Chemical Society removed this datum from the Journal's masthead some years ago).

Bancroft Rule States that the **external phase** of an **emulsion** will be the one in which the **emulsifier** is most soluble, e.g., sodium and alkali metal soaps, which are quite soluble in water, should lead to O/W emulsions. The original statement of the rule is more complex and depends on a model of the adsorbed film of **surfactant** as a duplex film, with *two* **interfacial tensions**, i.e., oil/film and water/film. The rule is, in fact, consistent with the approach based on **HLB number**.

11

Bashforth, Francis (1819-1912) British scientist. Author (with **J. C. Adams**) of *An Attempt to Test the Theories of Capillary Action* (Cambridge, 1883). Also wrote on theoretical ballistics.

Becher, Paul (b. 1918) American chemist. Author of numerous books and papers on **emulsions**.

Beer-Lambert law Governs the absorption or scattering of light, and is given by

$$I/I_0 = \exp(-\epsilon \Delta x)$$

where I and I_0 are the transmitted and incident light intensities, respectively, Δx is the length of the optical path, and ϵ is the *absorbance* or *extinction coefficient*. The Beer-Lambert law also applies to turbidity, where I is now the intensity of scattered light at some angle, and the absorbance is replaced by the **scattering** factor τ. Named for Wilhelm Beer (1797-1850), German astronomer, and Johann Lambert (1728-1777), German mathematician.[*]

Beilby layer The surface layer produced on a solid by polishing, which appears to be amorphous under the microscope.

BET *acronym* *B*runauer-*E*mmett-*T*eller. See **BET isotherm**.

BET isotherm An **adsorption** isotherm which takes into account the possibility of multilayer adsorption, given by

---

[*] Wilhelm Beer was the half-brother of the composer Giacomo (Jacob) Meyerbeer.

$$v/v_m = cx/(1 - x)[1 - (c - 1)x]$$

where $x = P/P^0$, and where v and v_m are, respectively, the volume adsorbed and the volume adsorbed at **monolayer** adsorption, P is the pressure of **adsorbate**, P^0 is its saturation pressure, and v is a constant related to the **partition function** of the adsorbate. Named for S. **Brunauer, P. H. Emmett**, and E. **Teller**.

bicontinuous *adj* A system consisting of at least two phases in which it is possible to move within either phase to any other part of that phase without crossing a phase barrier, i.e., an **interface**. An open-cell sponge is an example of a bicontinous system. It is probable that **emulsion inversion** proceeds by an intermediate bicontinuous structure.

bilayer *n* A two-dimensional membrane, separating two aqueous phases, made up of a twin layer of **surfactant** molecules aligned head-to-head, so that the outer surface is **lipophilic**. See **vesicle**.

Bingham, Eugene Cook (1878-1946). American chemist, founder of Society of Rheology.

Bingham flow Fluid behavior in which a critical shear is required before flow begins. See **Bingham, E. C.; yield value**.

Bingham plastic A substance which exhibits **Bingham flow**, i.e., has a **yield value**.

bimolecular film   See **bilayer**.

biocolloid *n*   A **colloid** substance of biological origin, e.g., **gums**, **lipids**, etc.

bipolar *adj* Characteristic of a substance or a surface which manifests *both* electron-acceptor and electron-donor functionalities. Because part of the **energy of cohesion** of such substances is due to the interaction between their electron-acceptor and electron-donor functionalities, the **surface tension** of bipolar substances is composed of a **Lifshitz-van der Waals** component and a **polar** component. Not to be confused with **dipolar**.

birefringence *n*   The property of exhibiting different refractive indexes in different directions, observed in certain crystals and in **liquid crystals**. Birefringence may be readily identified using polarized light, as with a polarizing microscope. —**birefringent** *adj*.

black lipid membrane   See **BLM**.

BLM *acronym* *B*lack *l*ipid *m*embrane. A thin film of a solution of organic lipids between two portions of aqueous solution. As the film drains, interference colors are first observed and then the film becomes black as it reaches **bilayer** thickness. See **cell membrane**.

block copolymer   A polymer composed of two monomeric species, in which the monomers polymerize in blocks, e.g.,

...AAAAABBBBBBAAAAAABBBBBBBB...

where A and B are the monomers. Block copolymers may exhibit **surfactant** properties.

Boltzmann, Ludwig (1844-1910) Austrian physicist. Noted for his work in the fundamentals of thermodynamics and gas kinetics.

Boltzmann constant The gas constant per molecule $k=R/N_A$, where N_A is the Avogadro number. See Appendix A.

Boltzmann equation Generally, any equation in which some property, such as potential, falls off exponentially with the distance. In colloid science, specifically the **Poisson-Boltzmann equation.**

bonding, hydrophobic See **hydrophobic bonding.**

Böttcher equation See Appendix E.

bottle neck pore See **ink bottle pore.**

Boyle point The temperature at which a non-ideal gas behaves ideally, e.g., obeys Boyle's Law. Analogous to the **theta temperature** in solutions of polymers.

Bragg equation The underlying equation for all diffraction phenomena

$$d \sin \Theta = \lambda/n$$

where d is the lattice spacing, Θ is the angle of the incident radiation, λ is its wavelength, and n is the refractive index of the medium. Named for Sir W. H. and Sir W. L. Bragg, British physicists, Nobel Prize 1915.

Bragg's law See **Bragg equation.**

breaking, of emulsions See **demulsification.**

Bredig arc A device for producing metal **sols,** by passing an electric current between two wires under the surface of a liquid. For Georg Bredig (1868-1944), German chemist.

Brønsted acid Any substance capable of donating a proton. On, e.g., a **catalyst,** this may simply be a lattice site. See **Brønsted base; Lewis acid.**

Brønsted base Any substance capable of accepting a proton. On, e.g., a catalyst, this may simply be a lattice site. See **Brønsted acid; Lewis base.**

Brønsted, Johannes (1879-1947). Danish physical chemist.

Brownian motion The erratic motion of small particles suspended in a liquid, readily observed under a microscope. It arise from random molecular collisions of the molecules of the suspending liquid (more accurately, from random fluctuations in the density of the liquid). Named for Robert Brown (1773-1858), British botanist, who first observed the effect with pollen grains in 1827. See **Langevin equation; random walk.**

Bruggeman equation See Appendix E.

Brunauer, Stephen (1903-1986). American chemist, born in Hungary. Noted for investigations of gas/solid adsorption. See **BET isotherm**.

bubble pressure, maximum See **maximum bubble pressure**.

builder *n* A substance, usually a pyrophosphate, added to a **detergent** formulation, to improve **detergency**. Its action is probably multiple; it acts to sequester ions contributing to hardness, it may affect surface charges (and thus stabilize soil particles), and it maintains a alkaline pH, which is required for optimum detergency. In recent years, concern for the ecological effects of the discharge of phosphate-containing detergents into surface waters have lead to a reduction in the quantity employed, and the use of other types of materials, e.g., sequestering agents, as a substitute.

Bungenberg de Jong, Henrik Gerard (1893-1977) Dutch chemist. Noted for his work on **coacervation**.

Burgers path An imaginary path on the surface of a crystal traced out by a clockwise circuit about some point, counting the same number of lattice points in the plus and minus directions. If the circuit does not close, a **dislocation** is present. The ends of the circuit define the **Burgers vector**. See **defect**.

Burgers vector The vector defined by a non-closing **Burgers path**, whose angle and magnitude define the type and magnitude of the **dislocation**.

C

Cabannes factor A correction term, used in **light scattering**, to allow for anisotropy.

carbon black Finely-divided form of carbon, whose large **surface area** makes it useful in, e.g., purification by **adsorption**. Also used as fillers, for example, in rubber.

calomel electrode Reference electrode, based on the redox reaction

$$Hg_2Cl_2(s) + 2e^- = 2Hg(l) + 2Cl^-(sat)$$

with a half-cell potential of 0.2415 volts if the electrolyte is saturated KCl. The name derives from the colloquial name, *calomel* for Hg_2Cl_2.

calorimeter *n* A device for the measurement of heat. See **calorimetry**.

calorimetry *n* Strictly, the measurement of heat; in practice, the measurement of heat evolved or absorbed during a chemical or physical change.

canal viscometer Device for the measurement of **surface viscosity**; essentially a two-dimensional analogue of the bulk **capillary viscometer**.

capillarity *n* A manifestation of **surface tension**, relating to **interfaces** which are sufficiently mobile to acquire an equilibrium shape, e.g., **menisci** and drops of liquid in air.

capillary *n* Of the dimensions of a hair (from L. *capilleus*); hence, a tube with a fine, hairlike, bore.

capillary action See **capillarity**.

capillary constant A dimensionless quantity defined as

$$a^2 = 2\gamma/\Delta\rho g$$

where γ is the **surface tension**, $\Delta\rho$ is the difference in densities of the phases involved, and g is acceleration due to gravity. See **capillarity**; **capillary rise**.

capillary condensation Of **adsorption** into capillaries or pores in a porous solid.

capillary electrometer Device for the measurement of **electrocapillarity**.

capillary pressure The stresses developed in a particle or droplet to balance the tension in the surface. See **Young-Laplace equation**.

capillary ripples Surface waves or ripples induced by some disturbance of the interface. Measurement of the damping of such ripples may supply useful information on the mechanical properties of **monolayers**. See **Rayleigh-Taylor instability**.

capillary rise A method of measuring **surface** or **interfacial tension** by observing the rise of a liquid in a **capillary**. In this case the **capillary constant** becomes rh where r is the capillary radius, and h is the rise. See **capillarity**.

capillary viscometer A device used for the measurement of viscosity, consisting of a U-tube, one side of which is a capillary, and the other side a reservoir. By measuring the rate of flow through the capillary arm, **Poiseuille's law** may be applied. Also **Ostwald viscometer**.

capillary waves Waves or ripples on the surface of a liquid, whose behavior is due more to **surface tension** than to gravity. Studies of the damping of such waves may provide information about the mechanical properties of surface films.

Casimir-Polder forces A modification of the **London** formulation, applying at distances very much greater than the wavelength. Named for H. B. G. Casimir and D. Polder, Dutch scientists.

catalysis *n* The causing or acceleration of a reaction by the action of a **catalyst**.

catalysis, micellar The causing or acceleration of a reaction in a **micellar solution**. The reactants are considered to be solubilized in a **micelle**.

catalyst *n* An agent which brings about or accelerates a reaction by increasing its rate or by increasing the yield of a particular product. Catalysts may be classified as **heterogeneous** (solid) or **homogeneous** (soluble). The classical definition of a catalyst was that of an agent which accelerated a reaction, but which did not take part in the reaction. This is not strictly true. See **enzyme**.

catalyst poison Additive which deactivates the catalyt surface, rendering it completely or partially ineffective.

catalyst promoter Additive which increases the activity of a catalyst surface.

cationic; cationic surfactant *n* An ionic **surface-active agent** (surfactant) in which the surface active moiety is the cation, e.g., cetyl trimethyl ammonium bromide $(CTAB)$, $C_{16}H_{33}N(CH_3)_3{}^+ Br^-$. See **anionic; nonionic**.

cathodic branch That portion of the **electrocapillarity** curve to the right of the maximum, i.e., the portion of the curve in which the mercury electrode is negatively charged. See **anodic branch**.

Cauchy plot A method of plotting refractive index n as a function of the frequency ω, in which n^2-1 is plotted against $\omega^2(n^2 - 1)$. The slope and intercept of the resulting straight line gives information about the ultraviolet behavior.

centrifugation n The separation of matter by means of an applied gravitational field, as in a *centrifuge*. See **ultracentrifuge** .

charge density The amount of charge per unit area on a surface. See **DLVO theory**.

cell membrane The outer layer of a biological cell. Such a membrane has many of the properties of a **monolayer**.

CFC *acronym* critical *f*locculation *c*oncentration. The minimum concentration of electrolyte which will induce **flocculation** of a sol. See *Schulze–Hardy rule*.

CFT *acronym* critical *f*locculation *t*emperature. The minimum temperature to which a **dispersion** must be raised to induce **flocculation**.

chaos theory The concept that the use of the calculus to describe natural phenomena may, in some cases, impose a regularity on the data which is in fact not present. If one, instead, employs an iteration process to describe a time series, at certain values of the parameters it appears that the system can exist in multiple random states. Although this concept has been applied, with some success, to, e.g., meteorology and economics, there is some question as to whether or not what is observed are artifacts of the calculation.

chemisorption *n* The process of **adsorption** characterized a by a chemical reaction between the **adsorbate** and **adsorbent**.

chi (χ) parameter The quantity in the **Flory-Huggins theory** which is a measure of solvent/solute interaction.

chi (χ) potential That part of the **surface potential** which arises from the presence of adsorbed dipoles.

chromatography *n* Any analytical procedure in which the components of a mixture are separated by flow through a packed column, e.g., *gas-liquid* chromatography. See **gel filtration; gel permeation; size exclusion.**

Clausius-Mossotti equation Defines the molar polarization *P* in terms of the relative **dielectric constant**

$$P = (M/\rho)(\epsilon - 1)/(\epsilon + 2)$$

where *M* is the molecular weight, ρ is the density, and ϵ is the relative dielectric constant. It should more properly be called the Mossotti-Clausius equation. Named for Rudolf Julius Emmanuel Clausius (1822-1888), German physicist and Ottaviano Fabrizio Mossotti (1791-1853), Italian physicist and mathematician.*

clay *n* Various forms of silicate minerals, e.g., kaolinite, pyrophilite, mica, montmorillonite, etc.

cloud point The temperature (usually Celsius) at which an aqueous solution of a **nonionic surface-active agent** separates into two phases, one of which is rich in the surface-active agent, and the other almost pure solvent. It is distinguished by the sudden onset of **turbidity** as the cloud-point temperature is reached. See **coacervation; Krafft point.**

cluster *n* In **catalysis.** an aggregate of a few (2 to about 12) metal atoms surrounded by various ligands, especially small attached molecules.

cmc; c.m.c. See **critical micelle concentration.**

coacervation *n* The separation of a solution (in a single solvent) containing two different **macromolecular** (or of one macromolecular and one micromolecular) species, of equal sign of charge (or of no charge), into two phases, one of which is rich in dissolved solute molecules of one species, while the other is enriched in the other dissolved solute. The phenomenon occurs in concentrated solutions or sols; **indifferent salts** have little influence; D.C. electric fields do not induce disintegration of the coacervate drops. The principal mechanism is reported to be phase separation engendered between **monopolar**

---

* Mossotti's name is frequently misspelled Mosotti in the literature and in textbooks.

(usually electron-donor) surfaces in a **bipolar** medium (e.g., water). See **complex coacervation; cloud point; Bungenberg de Jong, H. G.**

coadsorption *n* The simultaneous **adsorption** of two or more species.

coagulation *n* To form a coagulum by the action of an additive or temperature. Usually synonymous with **flocculation,** but may occasionally be used in a slightly different sense to mean a *dense* structure, as opposed to a more open one, as in a **floc.** If so used, this distinction should be clearly stated. —**coagulate** *v.*

coagulum *n* A loose structure formed of primary particles, **aggregates,** or **agglomerates** under the influence of weak attractive forces. See **secondary minimum.**

coalescence *n* The process whereby two (or more) small particles or droplets fuse together to form a single larger particle. —**coalesce** *v.* See **coagulate; flocculate.**

cohesion, energy of The intermolecular energy of attraction between the separate molecules constituting a homogeneous liquid or solid. This energy is made up of contributions from **Lifshitz-van der Waals** components in all cases; from **polar** components (in polar liquids or solvents); and from metal components in the case of metals. For liquids, it may be estimated as equal to twice the **surface tension**

$$W_C = 2\gamma$$

cohesive energy density A term coined by **J. H. Hildebrand**, used in the theory of imperfect solutions. The cohesive energy density is defined

$$c.e.d = \Delta E_v / [V]$$

where ΔE_v is the *energy* of vaporization, and $[V]$ is the molar volume. See **solubility parameter**.

coil, random See **random coil**.

co-ions *n* Ions of low relative molecular mass with a charge the same as the **colloidal electrolyte**. See **counter-ions**.

collapse pressure The **film pressure** at which a **monolayer** film begins to buckle, owing to the fact that the area is now smaller than is required for a close-packed film.

collector *n* In **flotation**, a **surface-active agent** adsorbed to the ore particle in order to render it sufficiently **hydrophobic**.

colligative properties Those properties of a substance which are dependent on their number rather than on its properties.

colloid *n* A particle in the size range $10^{-8} - 10^{-6}$ m, or, roughly in the range from molecular size to that visible under high magnification. The definition must be regarded as rather arbitrary. —**colloidal** *adj*. See **Graham, Thomas**.

colloidal electrolyte Used generally to describe an ionic **surface-active agent**, especially in the form of **micelles**.

colloid mill Device for preparing fine **dispersions** or **emulsions**, in which the material is sheared between a rotor and fixed stator. The clearance between the rotor and stator is adjustable, and usually quite small.

combinatorial formula In **adsorption** theory, the number of ways g_N in which in N molecules can be placed on S sites is

$$g_N = S!/N!(S-N)!$$

As in probability theory, where this equation gives the number of combinations of S objects taken N at a time.

comminution *n* The subdivision of particles by mechanical action, e.g., crushing.

complex coacervation The separation of a solution (in a single solvent) containing two different **macromolecular** (or of one macromolecular and one micromolecular) species, of different signs of charge, into two phases, one of which is rich in dissolved complexes formed between solutes of both species, while the other phase is depleted in solute molecules of both species. However, the concentration *ratio* of the solutes of the two species remains unchanged in both phases. See **coacervation; cloud point; Bungenberg de Jong, H. G.**

complex refractive index The representation of the **refractive index** of an absorbing substance by means of a complex number.

component *n* One of the set of the *minimum* number of constituents by which every **phase** of a given system may be described. Note that the number of components may be (and usually is) less than the number of *constituents*. See **phase rule**.

composite *n* A form of plastic in which fine particles or fibers, e.g., carbon fibers, are embedded to increase the strength of the material. The surface properties of reinforcing material are significant in determining the strength of the composite.

composite isotherm See **surface excess isotherm**.

compressibility *n* The reciprocal of the bulk modulus, equal to the ratio of the fractional change in volume to the stress applied.

compressibility, film The two-dimensional analogue of solid compressibility, given by

$$C^S = -(1/A)(dA/d\gamma)_T = (1/A)(dA/d\Pi)_T$$

where A is the area of the film, Π is the surface pressure defined by $\gamma_0 - \gamma$, where γ_0 is the surface tension of the substrate in the absence of the film, and γ is the surface tension with a film present.

concentration profile At a liquid/liquid or liquid/gas **interface**, the manner in which the concentrations of the various components (especially **surface-active agents**) varies in the direction normal to the interface. See **dividing surface**.

condensation *n* The association of primary particles to form larger ones. —**condensate** *n* That which has condensed. See **coagulation; flocculation.**

condensed state Of **monolayers,** a region of low compressibility, in which the monolayer is considered to have the properties of a two-dimensional liquid.

constitutive equations A set of integral equations which, in principal, contain all the information, for example, about an imposed simple shear stress and the resulting strain.

contact angle The angle formed at the line of contact at the interface between two liquids, a liquid, a solid and, a gas, or two liquids and a solid. By convention, the contact angle of a liquid is the *interior* angle in the liquid phase. See **advancing contact angle; receding contact angle; contact angle hysteresis.**

contact angle hysteresis The difference between the **advancing** and **receding contact angle.**

continuous phase In a **dispersion** or **emulsion,** the **phase** in which the particles or droplets are dispersed. See **discontinuous phase.**

copolymer *n* A polymer, composed of two or more monomer types. The monomers may be randomly disposed, or in the form of blocks. See **block copolymer.**

corrosion *n* The act of eating away gradually, as if by gnawing, especially by chemical action, usually at a metal surface. Also the product of corrosion. From L. *corrodere* to gnaw to pieces. —**corrode** *v.* See **passivation**.

cosurfactant *n* A low-molecular-weight **surfactant**, e.g., a lower fatty alcohol, which acts in conjunction with a normal surfactant to form a **microemulsion.**[*] See **hydrotrope**.

couette flow Flow of a fluid between two surfaces that have tangential relative motion, e.g., two coaxial cylinders rotating at different speeds. See **couette viscometer; laminar flow.**

couette viscometer A device for the measurement of viscosity, in which the response of a suspended cylinder to liquid flow induced by the rotation of a larger, concentric cylinder is determined. From Fr. *couette*, lit. a featherbed, but, in machinery, a bearing.

Coulomb's law The force F acting between two charges in a medium is given by

$$F = q_1 q_2 / 4\pi\epsilon_0 \epsilon r^2$$

where q_1 and q_2 are the charges, ϵ_0 and ϵ are the permittivities of free space and of the medium, respectively, and r is the distance of separation. Named for Charles Augustin de Coulomb (1736–1806), French physicist.

counter-ion; counterion *n* Ions of low relative molecular mass,

---

[*] Sometimes used, loosely, to describe the second member of an emulsifier pair. Although this is not strictly incorrect, the usage should be restricted to the microemulsion situation.

with a charge opposite to that of the colloidal ion. See **co-ions; gegenions.**

creaming *n* The phenomenon of *upward* **sedimentation,** occurring when the density difference between a suspended particle and the suspending liquid is negative. So-named by analogy with the separation of cream in unhomogenized milk. —**to cream** *v.* See **Stokes' law.**

creep *n* The flow of a **viscoelastic** material.

creep compliance The response of a **viscoelastic** material to an applied shear.

critical flocculation concentration See **CFC.**

critical flocculation temperature See **CFT.**

critical micelle concentration The concentration, in solution, at which a **surface-active agent** forms multimolecular aggregates, which are in kinetic equilibrium with monomer (i.e., single) molecules. The phenomenon is accompanied by an abrupt change in certain physical properties, the most notable of which is the break in the surface or interfacial tension curve. Commonly abbreviated **cmc.** More accurately, *critical concentration for micelle formation.* See **micelle.**

critical surface tension The surface tension above which a liquid cannot completely wet a solid, e.g., **contact angle** > 0. The critical surface tension is sometimes taken to be the **surface tension** of the solid. Coined by W. A. Zisman (1905-

1986).

critical temperature, two-dimensional Of a monolayer, the temperature above which the compression of the monolayer does not exhibit the presence of a two-dimensional liquid region. In gas-solid **adsorption**, the critical temperature of the adsorbed layer of gas, usually substantially lower than the bulk (three-dimensional) value.

cross section For **absorption** and **scattering**. In **light scattering**, the effective cross-section of the absorbing or scattering particle. It is not the physical cross-section, but rather a measure of the *blocking power* of the particle as far as the transmission of light is concerned.

crystal defects See **defects, crystal**.

cumulative distribution A distribution function in which the distribution of, e.g., particle sizes, is *cumulative*, i.e., the total number or number fraction of particles above (or below) a given size. See Appendix C.

Curie, Pierre (1859-1906) French physicist. Codiscoverer of radium. Nobel prize in physics in 1903.

Curie point The temperature above which a **ferromagnetic** substance becomes **paramagnetic**. Also **Curie temperature**.

current, streaming See **streaming current**.

D

Darcy's Law   For the flow of a viscous liquid in a porous medium the medium is treated as a bundle of capillaries, the flow rate Q being given by

$$Q = K(A\Delta P/\eta l)$$

where A is the total cross-sectional area of the porous medium, l is its length, ΔP is the pressure drop, and η is the viscosity of the fluid. The quantity K is the *permeability* of the medium.

dark field microscopy   A technique in optical microscopy for the examination of particles, especially in **suspension**, in which the particles are illuminated from the side, rather than from below.   By this technique, particles as small as 5 nm may be detected, with, however, extensive loss of detail. This technique is useful for counting or tracking particles, as, e.g., in **electrophoresis** measurements.

dead space In manometric measurements, e.g., as in the determination of gas **adsorption**, a volume of the equipment which, for mechanical reasons, is not occupied by sample, and must be corrected for.

Deborah number In **rheology**, a dimensionless quantity D_N defined as the ratio of the **relaxation time** of the material under study to the time of observation. When $D_N \to 0$ the materials tend to behave like fluids; when $D_N \to \infty$ they behave like solids. Coined by **M. Reiner**, from Judges 5:5, which may be translated "The mountains flowed down...".

Debye, Peter Joseph William (1884-1966) Dutch physical chemist, noted for the theory of strong electrolytes, with E. A. A. J. Hückel (1896-1980) and for many contributions to colloid science. Awarded Nobel Prize in chemistry in 1936.

Debye forces The **interaction energy** between a dipolar and a neutral molecule, arising from a **dipole** induced in the neutral molecule, and proportional to the inverse sixth power of the distance.

Debye-Hückel theory A theory of the behavior of strong electrolytes in solution, involving the solution of the **Poisson-Boltzmann** equation for the potential of a point charge. A version of this solution is the basis for the **DLVO Theory.**

Debye length A quantity found in modern theories of ionic solutions, equal to $1/\kappa$, where κ is defined by

$$\kappa^2 = e^2 \Sigma_i n_i z_i / \epsilon kT$$

where e is the electronic charge, ϵ the **permittivity**, n_i the number of ions i per unit volume, and z_i their valence. For dilute aqueous solutions at $25°$

$$\kappa = 3.29 \times 10^9 \, c^{1/2} \, m^{-1}$$

where c is the molarity of the solution. Also **Debye parameter**. See **DLVO Theory**.

Debye parameter; Debye-Hückel parameter See **Debye length**.

deconvolution n The mathematical process by which a multimodal distribution is broken down to its underlaying distributions, i.e., *deconvoluted*.

defect n An imperfection in a crystal lattice. Two important defects are the so-called **Schottky** and **Frenkel defects**. Defects are sometimes characterized as *kinks*, *steps*, or *terraces*, which designations are sufficiently descriptive.

degrees of freedom The number of *independent* variables required to define a system. See **phase rule**.

degree of polymerization The number of repeat (monomer) units in a polymer chain. Frequently abbreviated DP. See **random coil**.

Democritus (c. 460-370 BCE) Greek philosopher, believed to be the originator of the concept of atoms. Known as the *Laughing Philosopher*.

demulsification *n* The breaking, or separation, of an **emulsion** into its separate phases. Demulsification may occur without intervention through **sedimentation**, **creaming**, or **coagulation**. Such emulsions are unstable. With stable emulsions, demulsification may be induced by the addition of suitable **demulsifiers**, by centrifugation, by the use of various types of electrical discharge, or by passing through a suitably packed column; in some cases, it may even be induced by gentle agitation. —**demulsify** *v*.

demulsifier *n* A substance, usually a **surface-active agent**, added to an emulsion to induce demulsification. A simple interpretation of their action is that they change the **HLB** of the surfactant present to one which is unsuitable for stabilizing the system. See **required HLB**.

depletion flocculation **Flocculation** induced by polymer in free solution, arising when the **colloidal** particles approach so closely that polymer chains are excluded from the interphase region.

depletion stabilization In effect, **stabilization** brought about by somewhat the same mechanism as **depletion flocculation**, i.e., the presence of polymer chains in free solution. The experimental evidence for this effect is somewhat doubtful.

depolarization *n* In **light scattering**, light from an isotropic particle scattered at 90° is totally polarized; the polarization is reduced for anisotropic particles, and the **depolarization ratio** can be used to calculate the **Cabannes factor.**

depolarization ratio In **light scattering,** the ratio of the intensity of the horizontally polarized light to that of the vertically polarized.

Derjaguin, Boris Vladimirovich (b. 1902) Soviet chemist. See **DLVO theory; water, anomalous.**

desorption *n* The removal of molecules or atoms from a surface; the antonym of **adsorption.** —**desorb** *v.*

detergent *n* A **surface-active agent** employed for removing dirt and soil, e.g., a *household* detergent.

deviation, standard See Appendix D.

dialysis *n* The process of separating low molecular weight impurities from a colloidal system by the use of a **semipermeable membrane.** See **osmosis.**

diameter, mean *n* See Appendix B.

dielectric constant See **permittivity.**

dielectric saturation The reduction in **dielectric constant** owing to the presence of an electric field. This becomes a problem in defining that quantity in the **Stern layer.**

differential capacitance The rate of change of charge density with voltage at an interface.

differential refractometer A device for measuring small differences in refractive index between, e.g., a solution and a reference solvent, permitting direct determination of the rate of change of refractive index with concentration (dn/dc), required for **light scattering** measurements.

diffuse double layer That portion of the **double layer** which is described by the **Poisson-Boltzmann equation**. See **DLVO Theory**.

diffusion n The migration of **colloid** particles, molecules, or ions under a concentration gradient. See **Fick's first law**; **diffusion coefficient**.

diffusion coefficient The constant coefficient in **Fick's first law**. For spherical particles the diffusion coefficient D is given by

$$D = kT/6\pi r\eta$$

where r is the particle radius, η is the viscosity of the medium in which the diffusion occurs, and k and T have their usual meanings. A measure of the ability to diffuse. See **diffusion**.

diffusiophoresis n The movement of a rigid **colloidal** particle owing to a gradient of molecular solute.

diffusivity See **diffusion coefficient**.

dilatant *adj* Of a system in which the viscosity increases with increase in the rate of shear in a time-independent way. —**dilatancy** *n* See **rheopexy**.

dipole n In *Physics*, a pair of electric point charges of equal magnitude and opposite sign, separated by an infinitesimal distance. In *Chemistry*, a polar molecule. —**dipolar** *adj*.

dipole-dipole interaction See **Keesom forces**.

dipole moment A vector quantity associated with a **dipole**, equal to the product of the charge and the distance of separation, and having a direction from the negative to the positive charge along the line between the charges.

discontinuous phase In a **dispersion** or **emulsion**, the phase which is dispersed as particles or droplets in the **continuous phase**.

dispersed phase See **discontinuous phase**.

dispersion forces See **London forces**.

discreteness of charge In **DLVO theory** the surface charge is treated as being smeared out, i.e., uniformly distributed over the surface of the **colloid** particle. However, since the source of the surface charge may be, in fact, discrete charges, this approach cannot be correct on a microscopic scale. By correcting for *discreteness of charge* a more reasonable value for the surface potential is obtained.

disjoining pressure   Term coined by **B. V. Derjaguin** to describe surface forces arising from overlapping interfacial layers. Initially restricted to *repulsive* forces, more recently the definition has been extended to include attractive forces. It thus comes close to embracing the total interfacial interaction, and therefore may be thought unnecessary.

dislocation *n*   Approximately, a concentration of **defects** in a crystal lattice. See **Burgers path; Burgers vector.**

dispersant *n* A **surfactant**, specifically one useful for promoting the dispersion of solids in a liquid.

disperse *v*   To cause particles (of one phase) to separate uniformly throughout another phase, i.e., gas, liquid or solid. —*adj* as in *disperse system.* In *Optics*, to subject light or other electromagnetic radiation to **dispersion.**

dispersing agent   See **dispersant.**

dispersion *n*   The act of **dispersing.** —*n* A **disperse** system. In *Optics*, the separation of electromagnetic radiation into its individual frequencies.

dispersion forces   See **London forces.**

displacement, mean square   See **root mean square displacement.**

dissipative process   Losses in a process arising from frictional or viscous effects.

dissymmetry ratio In **light scattering**, the ratio of the light scattered at 45° to that scattered at 135°. This ratio is ultimately related to the **radius of gyration** of the scattering body, and can thus be used to determine molecular size and shape. —Also **dissymmetry** *n*.

distribution function See Appendix D.

dividing surface An imaginary surface, defined by J. W. Gibbs, which separates the bulk from the **surface phase**. It is a mathematical device, permitting the statement of separate thermodynamic functions for the surface and the bulk. See **concentration profile**.

DLVO theory *acronym* *D*erjaguin-*L*andau-*V*erwey-*O*verbeek. The theory of colloid stability arising from electrostatic repulsions (based on the solution of the **Poisson-Boltzmann equation** with the appropriate boundary conditions) balanced against attractive potentials arising from **Lifshitz-van der Waals forces.** See **double layer.**

Donnan equilibrium For electrolyte separated by a membrane, the **mean ionic activity** is equal on both sides.

Donnan, Frederick George (1870-1956). British chemist.

Doppler broadening The increase or decrease in frequency of radiation, e.g., scattered light, depending on whether the source of the radiation is moving toward or away from the observer. —Also **Doppler effect; Doppler shift.** For C. J. Doppler (1803-1853), Austrian physicist.

double layer; double layer of charge The duplex layer of charge existing around the surface of a **micelle**, a polyelectrolyte molecule, or a solid, consisting of a charge on the surface of the particle, and of the ions of opposite charge (**gegenions** or **counterions**) concentrated near the surface to ensure local electrical neutrality. The charge on the surface of a particle may be an inherent property of the solid (e.g., in ionic crystals) or of the polyelectrolyte, or may be created by the **adsorption** of **surface-active** molecules. The double layer may be divided into the **Stern layer** (close to the particle) and the **Gouy-Chapman layer**, extending into the dispersion medium. See **DLVO theory**.

Dougherty-Krieger equation See **Krieger-Daugherty equation**.

drag *n* Frictional force on a particle in motion arising from the viscosity of the suspending medium. For a spherical particle it is equal to $6\pi r\eta u_t$ where *r* is the radius of the sphere, η is the viscosity of the suspending medium, and u_t is the terminal velocity of the particle.

droplet size distribution See **particle size distribution**.

dropping mercury electrode A device for measurements at the mercury/solution **interface**, in which the surface of the mercury is constantly renewed by falling in drops from a capillary. The dropping mercury electrode is the basis of the polarograph, used for analytical purposes. See **electrocapillarity**.

drop weight method Method for the determination of **surface tension**, in which a drop (or a number of drops) are allowed to form at the tip of capillary and fall under the effect of gravity. The weight of the drop is then given by

$$W = 2\pi r\gamma$$

where r is the *external* radius of the capillary, and γ is the surface tension. For the method to work properly, the end of the capillary must be exactly plane and highly polished. For exact measurements, the corrections of **Harkins** and Brown must be used; these tabulated values depend on R, the *internal* radius of the capillary a, and the volume of the drop V.

duNoüy tensiometer A device for the measurement of **surface and interfacial tension**, in which the force required to pull a platinum-iridium ring through the surface or interface is measured by means of a torsion balance. The weight measured is the sum of the weight of the ring W_{ring} plus the detachment force

$$W_{tot} = W_{ring} + 4\pi R\gamma$$

where R is the *internal* radius of the ring and γ is the surface or interfacial tension. For exact measurements the corrections of **Harkins** and Jordan must be applied; these tabulated values depend on two dimensionless quantities R^3/V and R/r, where V is the volume of the meniscus and r is the radius of the platinum-iridium wire. For Pierre LeComte duNoüy (1883-1947), French-American biochemist.

duplex film See **film, duplex.**

Dupré, Athanase Louis Victoire (1808-1869) French physicist.

Dupré equation See **adsorption, energy of.**

dynamic light scattering See **PCS.**

E

eccentricity *n* Departure from sphericity; for an ellipsoid it is equal to $(a-b)/b$, where a is the length of the major axis, and b that of the minor.

eigenfunction *n* In mathematics, the set of discrete solutions to a second-order differential equation. The solutions of the Schrödinger equation of quantum mechanics are eigenfunctions of the quantum numbers. From Ger. *eigenfunktion* particular (or individual) function.

Eilers equation See Appendix F.

Einstein, Albert (1879-1955) German-American physicist. Although most famous for his Theory of Relativity, he made important contributions to colloid science, through the theory of the **Brownian motion** (1905) and of the **rheology** of **disperse systems** (1906). Nobel Prize in Physics in 1922.

Einstein equation, for diffusion   The diffusion coefficient D of a particle is given by

$$D = kT/f$$

where f is the **friction factor**, equal to $6\pi\eta r$

Einstein equation, for viscosity   For a dilute dispersion of rigid spheres, **Einstein** showed, on hydrodynamic grounds, that the viscosity is given by

$$\eta = \eta_o(1 + 2.5\phi)$$

where η is the viscosity of the suspension, η_o is the viscosity of the dispersion medium, and ϕ is the **phase volume** of the dispersed spheres. In spite of the severe limits, i.e., dilute solution, rigid spheres, this equation is valid in a surprising number of cases. To get around the limits, a number of variations have been suggested; see Appendix F.

Einstein-Smoluchowski equation   The root-mean-square displacement of a particle undergoing **Brownian motion** is given by

$$<x>^{1/2} = (2Dt)^{1/2}$$

where D is the **diffusion coefficient** and t is the time.   See

Einstein equation, for diffusion.

elasticity *n* The property of a substance that enables it to change its length, volume, or shape under an applied force, and return to its original state on removal of the force. See **elastic limit; modulus of elasticity; surface elasticity.**

elastic limit The greatest stress which can be applied to an elastic body without permanent deformation.

electric double layer See **DLVO theory.**

electric potential *n* At any point in an electric field, the work required per unit charge in moving an infinitesimal point charge from a common reference point (usually infinity) to the given point. See **Galvani potential; Volta potential.**

electrocapillary maximum The point in the **electrocapillarity** curve at which $\partial\gamma/\partial E = 0$. See **zero point of charge.**

electrocapillarity *n* The study of the change in shape (and hence **interfacial tension**) of a mercury droplet in an electrolyte solution under an applied electric field. Such studies throw considerable light on the structure of the **electrical double layer.** First studied by G. Lippmann in 1875.

electrochemical potential Analogous to the chemical potential, it arises when a charge is moved from one phase to another. As with the chemical potential, the electrochemical potential must be the same in all phases at equilibrium. See **chi potential; Galvani potential; Gibbs-Duhem equation.**

electrocratic *adj* Denotes a dispersed-phase system stabilized by electrostatic repulsion. Coined by E. A. Hauser (1896-1956).

electrode, calomel half-cell See **calomel electrode**.

electrode, reference An electrode of known potential (e.g., a **calomel electrode**), used in determining the potential of an electrochemical cell.

electrodialysis *n* The process of enhancing the **dialysis** of ionic species by placing electrodes in the compartment surrounding the enclosed **colloid** and taking advantage of the migration of ions in an electric field. See **osmosis**.

electrokinetics *n* The study of the motion of charged particles in an electric field. See **electrophoresis; zeta potential**.

electrokinetic potential See **zeta potential**.

electrolyte, colloidal See **colloidal electrolyte**.

electron acceptor An atom or molecule which *accepts* an electron pair from an **electron donor** to form a covalent bond.

electron donor An atom or molecule which *donates* an electron pair to an **electron acceptor** to form a covalent bond.

electron microscope, scattering See **SEM**.

electron spin resonance See **ESR**.

electro-osmosis *n* A method for the measurement of zeta potential, in which the electrode compartments are separated by a capillary or porous plug. Also, in general, the flow of a liquid through a semipermeable membrane under an electric field. See electrophoresis.

electrophoresis *n* The motion of an ion or a colloid particle under an applied D.C. field. See electrophoretic mobility; zeta potential.

electrophoretic mobility *n* The velocity of motion of an ion or particle undergoing electrophoresis. In the simplest case, the velocity is given by the Helmholtz-Smoluchowski equation

$$v = \zeta E/4\pi\eta$$

where ζ is the zeta potential, E is the applied D.C. drop, and λ is the viscosity of the suspension medium. However, the velocity is also a function of the size and shape of the drop, and correction terms must be applied in precise measurements (see Henry equation).

electrosteric stabilization A combination of steric and electrolytic (double layer) stabilization imparted by the adsorption of a polyelectrolyte.

electroviscous effect The increase in viscosity of a disperse system arising from a charge on the particles. Smoluchowski modifed the Einstein equation to take this into account

$$(\eta - \eta_0)/\eta_0 = 2.5\phi[1 + (\epsilon\zeta)^2/4\pi^2\eta_0\kappa a^2]$$

where a is the radius of the particle, κ is the specific conductivity of the dispersion, ϵ and η_0 are, respectively, the **dielectric constant** and viscosity of the dispersion medium, and ζ is the **zeta potential** of the charged particles.

ellipsoid *n* Three-dimensional figure obtained by the rotation of ellipse about either of its major axes. If the rotation is about the longer axis, the ellipsoid is **prolate**; if about the shorter axis, it is **oblate**.

ellipsometry *n* The measurement of the elliptically polarized portion of light reflected from a surface at the Brewsterian angle ($\tan^{-1} n$, where n is the refractive index of the substrate). Applied to **monolayers**, such measurements give useful information about the orientation of molecules at the **interface** and of the thickness of the surface layer. For solids, ellipsometry may be used in the investigation of corrosion.

Emmett, Paul Hugh (1900-1985) American chemist, noted for his investigations of gas/solid adsorption. See **BET isotherm**.

emulsification *n* The process of making an emulsion. —**emulsify** *v*.

emulsifier *n* A substance used to stabilize an **emulsion**; usually a **surface-active agent**, but also may be a polymer or a finely-divided solid.

emulsion *n* A dispersion of one immiscible liquid in another, in the form of fine droplets, approximately in the range of 100-1000 nm. Such dispersions possess minimal stability, which may be enhanced by the interfacial adsorption of a third component. The third component may be a **surface-active agent** (or a mixture of surface-active agents), a finely-divided solid, or a surface-active polymer (either synthetic or natural). The phases making up an emulsion are traditionally referred to as oil and water, but may be any two liquids differing markedly in polarity. Two emulsion types are recognized: oil-in-water (O/W) and water-in-oil (W/O). The latter is sometimes referred to as an *invert* emulsion. In recent years, it has been convenient to use the term **macroemulsion** to distinguish between ordinary emulsions and **microemulsions**. See **miniemulsion**.

end effect In viscometry, the fact that the flow at the ends of, e.g., a **couette viscometer**, is different from that at other points. In precise work, a correction for the end effect must be applied.

enthalpic stabilization See **steric stabilization**.

entropic stabilization See **steric stabilization**.

enzyme *n* Any of a class of naturally-occurring proteins capable of functioning as a **catalyst** in reactions of organic substances. From Gk *enzymos* leavened. —**enzymic** *adj*.

EOR *acronym* For *E*nhanced *O*il *R*ecovery. See **oil recovery**.

Eötvös equation For the temperature dependence of the surface tension of a liquid

$$\gamma^{2/3} = k(T_c - T)$$

where V is the molar volume, T_c is the critical temperature of the liquid, and T is the temperature of the measurement. k has the value of 2.1 for most liquids. The equation was modified by W. Ramsay (1852-1916), English chemist, and J. Shields by substituting $(T_c - 6)$ for T_c in the above equation. For Roland Eötvös (1848-1919), Hungarian physicist.

epitaxy *n* An oriented overgrowth of crystalline material on the surface of a crystal of different, but similar, structure. —**epitaxial** *adj*.

EPR See ESR.

equation, named There are a large number of named equations of importance in colloid and interface science. Such equations are usually listed in the Dictionary under the name of the scientist who is credited with their derivation.

equilibrium, Donnan See **Donnan equilibrium**.

equilibrium, thermodynamic The condition under which the **chemical potential** of any component of a system must be equal in all parts of the system to which the component has access.[*]

Esin-Markov coefficient Measures the effect of electrolyte on the **electrocapillary maximum** or the **zero point of charge**.

---

[*] This definition lies at the heart of many important relations, e.g., the **Gibbs-Duhem equation**.

ESR *acronym* electron *s*pin *r*esonance. Also **electron paramagnetic resonance; EPR.** Measurement of the flipping of the sign of electron spin in a magnetic field of the correct frequency. Such measurements give useful information about molecular structure and orientation. See **NMR.**

excess concentration See **surface excess.**

excluded volume The volume of a system which is unavailable for a molecule or particle owing to the presence of other molecules or particles. The *available* volume is then the total volume less the excluded volume.

expansion factor The ratio of the **radius of gyration** of a molecular coil to that which would obtain in the case of zero interaction with the solvent. See **random coil; theta temperature.**

extinction coefficient See **Beer-Lambert law.**

F

Faraday constant See Appendix A.

FEM *acronym* *f*ield *e*mission *m*icroscopy. In FEM, electrons emitted from a finely pointed and charged metal tip travel in straight lines to a cathode ray screen, which acts as an anode. The crystal planes making up the tip can be seen, with a resolution of 300-500 nm.

Feret's diameter The distance between two tangents on opposite sides of a particle, and parallel to some fixed direction.

ferromagnetic *adj* Of a material which, below the **Curie point,** possesses magnetic properties in the absence of an electric field. **—ferromagnet** *n.* See **paramagnetic.**

Fick, Adolf Eugen (1829-1901) German physiologist.

Fick's first law For the rate of diffusion under a concentration gradient

$$J_i = -D(dc/dx)$$

where J_i is the flux of the ith species, D is the **diffusion coefficient**, and dc/dx is the gradient. For **A. E. Fick**.

Fick's second law For the time rate of change of the concentration gradient

$$\partial c/\partial t = D[\partial^2 c/\partial x^2]$$

where D is the **diffusion coefficient** and c is the concentration at time t and location x. For **A. E. Fick**.

field emission microscopy See **FEM**.

field ion microscopy See **FIM**.

filler n A finely-divided material added to paper, paint, rubber, etc., to impart additional strength. See **composite**.

film balance An apparatus for the preparation and manipulation of **monolayers**. In the form popularized by **I. Langmuir**, it consists of a long, shallow trough filled with high purity water (or, in certain cases, aqueous solutions) on which is spread a monolayer, with a float system for measuring **surface pressure**. The surface pressure may also be measured by means of a **Wilhelmy plate**. The monolayer is compressed or expanded by means of flat barriers. The film balance provides basic information on molecular geometry and orientation, location and strength of polar groups, and forces of cohesion and adhesion. See monolayer; Pockels, Agnes.

film, black See BLM; foam drainage.

film, built-up See Langmuir-Blodgett film.

film drainage The drainage of interlamellar liquid from a single detergent film. As the film becomes thinner, interference colors appear and travel down the film, permitting measurement of the film thickness as a function of time. If the film thickness becomes less than wavelength of the light illuminating the film, a **black film** results. See **foam drainage**.

film, duplex In an **emulsion**, the **monolayer** of **surfactant** surrounding the dispersed droplet can be regarded as possessing a surfactant/water **interfacial tension** on one side, and a surfactant/oil interfacial tension on the other, forming a *duplex* film. The curvature of the interface then depends on the relative values of the two interfacial tensions, and thus determines whether the emulsion is O/W or W/O. In effect, this is the **Bancroft rule**. See **oriented wedge**.

film elasticity See **surface elasticity**.

film pressure The two-dimensional pressure exerted by an adsorbed **monolayer**

$$\Pi = \gamma_{solvent} - \gamma_{solution}$$

where γ is the **surface tension**. The film pressure may be measured directly in **film balance**.

FIM *acronym* *f*ield *i*on *m*icroscopy. Similar to FEM, except that the tip is positively charged. As a result, a gas molecule (e.g., helium) which approaches the tip is stripped of an electron, and the resulting positive ion moves radially to the negatively charged screen. The resulting image has much greater resolution than is possible with **FEM**, and individual atoms on the tip may be seen.

floc *n* A loose structure formed of primary particles, **aggregates**, or **agglomerates** under the influence of weak attractive forces. See **secondary minimum**.

flocculation *n* To create **flocs** by the addition of a suitable agent or by a change of physical conditions. —**flocculate** *v*.

flocculation, depletion See **depletion flocculation**.

flocculation value The concentration of electrolyte required to coagulate a sol in a given time interval. See **Schulze-Hardy rule**.

Flory, Paul John (1910-1985) American chemist, noted for work in the field of polymers. Nobel prize 1974.

Flory-Huggins theory A theory of polymer solutions in which a polymer solution is envisioned as a three-dimensional lattice. A lattice site is able to accommodate either a solvent molecule or a polymer segment. Named for **P. J. Flory** and **M. L. Huggins**.

Flory-Krigbaum theory A theory of polymer solution which takes into account the dimensions of the polymer coils and the effect of excluded volume. Named for **P. J. Flory** and **W. R. Krigbaum.** See **Flory-Huggins theory; steric stabilization; theta temperature.**

Flory point See **theta temperature.**

Flory temperature See **theta temperature.**

flotation *n* In general, a process for the separation of solids by agglomerating the desired solid particles into a **foam** or **froth**, and separating the foam from the mass of liquid. In mineral beneficiation, the use of appropriate **surface-active agents** (which may control the liquid/solid **contact angle**) makes it possible to separate a desired mineral species from a mixture, e.g., gold from quartz, by froth flotation.

fluctuation *n* A random (positive or negative) change in some property of a system. Fluctuations in density are, for example, of importance in the analysis of **Brownian motion** and **light scattering.** —**fluctuate** *v.*

fluid dynamics That portion of **fluid mechanics** dealing with the properties of fluids in motion. See **hydrodynamics.**

fluid mechanics Applied science dealing with the basic properties of liquids and gases. See **fluid dynamics.**

flux *n* The rate of flow of fluid, particles, or energy; a quantity expressing the strength of field of force in a given area.

foam *n* A coarse dispersion of gas in a liquid, which may be stabilized by a **surface-active agent**. See **froth; kugelschaum; polyederschaum**.

foam drainage The separation of liquid from the body of the foam. In coarse foams (**kugelschaum**) the drainage is governed by hydrodynamic factors, indeed a form of the **Poiseuille equation** may be employed. In more concentrated foams, where the bubbles are in the form of polyhedra (**polyederschaum**), and where a **Plateau border** is formed at the intersection of three bubbles, the foam drains from the lamellae, resulting in lamellar thinning, which may contribute to foam instability. See **film drainage**.

fog *n* A dispersion of water-droplets in air. The opacity of fog is due to the **light scattering** from the colloidal droplets. By extension, any liquid/gas **dispersion**.

Fowkes equation For the calculation of **interfacial tension**

$$\gamma_{12} = \gamma_1 + \gamma_2 + 2(\gamma_1^{\,d}\gamma_2^{\,d})^{1/2}$$

where the γ^d are effective surface tensions attributable only to the **dispersion** component of the **surface tension**, and can, in principle, be calculated from molecular properties, or determined from appropriate measurements, the basic assumption here being that the surface tension of linear hydrocarbons consists solely of the dispersion contribution. Named for **Frederick M. Fowkes**. See **Antonoff rule; Good-Girifalco equation**.

Fowkes, Frederick Mayhew (b. 1915). American chemist. See

Fowkes equation.

Fowler-Guggenheim equation A version of the **Langmuir isotherm** which takes into account the possibility of lateral interactions between the adsorbed molecules

$$P = K(\theta/1 - \theta) \exp(\epsilon\theta/kT)$$

where P is the pressure, θ is the fractional surface coverage, and ϵ is the lateral interaction energy.

fractal *n* A structure having an irregular or fragmented shape at all scales of measurement. Coined by Benoit Mandelbrot (b. 1924), French-American mathematician.

fractal dimension The exponent relating the mass or surface of a body to its size

$$M \propto R^{D}$$

where M is the mass (or volume), R is some characteristic dimension of the body (e.g., the radius for a sphere), and D is the fractal dimension. For a sphere, of course, R is the radius and $D = 3$, but one may define bodies for which D is not necessarily an integer. For example, colloidal **aggregates** of silica may have fractal dimensions in the range 1.80-3.00, depending on the kinetics of the aggregation.

Franklin, Benjamin (1706-1790) American statesman, scientist, and philosopher. The first to estimate the thickness of a **monolayer**, he used his observations to explain the calming effect of oil on waves (i.e., *pouring oil on troubled waters*).

free-draining *adj* Of a polymer coil in the extended or unwound state, so that streamline flow occurs. See **non-draining**.

free volume In polymer solution, that portion of the total volume not occupied by the geometrical volumes of the constituent molecules. Cf. **excluded volume**.

free polymer In a polymer-stabilized **dispersion**, any polymer which is not adsorbed to a colloidal particle, i.e., in free solution. See **depletion flocculation; depletion stabilization**.

freeze fracture technique A method of preparing specimens for **electron microscopy** by rapid freezing, cleaving the frozen specimen with a sharp knife or razor, and coating the specimen with a thin layer of metal by sputtering.

Frenkel defect **Defect** in a crystal lattice arising from the migration of some ions (usually cations) to interstitial positions, thus leaving lattice vacancies.

Frenkel-Halsey-Hill isotherm An isotherm for multilayer adsorption on porous solids, somewhat similar to the **Polanyi adsorption isotherm**.

Freundlich, Herbert M. F. (1880-1941) German-American physical chemist. Author of *Colloid and Capillary Chemistry* (1926).

Freundlich isotherm For **adsorption** from solution, given by

$$\Theta = ac^{1/n}$$

where Θ is the fraction of the **adsorbent** surface covered, *c* is the equilibrium concentration of the **adsorbate**, and *a* and *n* are constants, with *n* > 1. Note that *a* and *n* are not arbitrary constants, but may in principle be calculated. As a practical matter, however, they are treated as empirical constants, to be determined by experiment. See **Langmuir isotherm**.

Fricke equation See Appendix E.

friction *n* **Surface** resistance to relative motion, e.g., a body sliding or rolling. In most cases, frictional resistance occurs because of the small-scale roughness of the surfaces in contact. —**frictional** *adj.* See **Amontons' law; lubrication.**

friction factor *n* In **rheology** of **dispersions**, a measure of the dissipation of energy due to internal friction. For spheres

$$f = 6\pi\eta r$$

where *r* is the radius of the sphere, and *η* is the viscosity of the dispersion medium. See **drag; Stokes' law.**

froth *n* A **foam**, usually one with open, lacy structure, e.g., sea foam.

frothing agent A **surfactant** capable of producing **foam** or **froth**. Specifically, one used in **froth flotation**.

froth flotation See **flotation**.

fuzzy set In mathematics, a set for which the concept of membership is "fuzzy," i.e., it is not always clear whether or not an object belongs in a set. This rather esoteric branch of set theory has found application in the description of particle shapes.

G

Galvani potential The electrostatic potential *within* a phase, i.e., $1/e$ times the work required to bring a unit charge from infinity into the phase. Also **inner potential**. Named for Luigi Galvani (1737-1798), Italian physiologist. See **Volta potential**.

gaseous monolayer See **ideal gas, two-dimensional**.

gegen-ion; gegenion *n* From Ger. *gegen* against, opposed to + ion. See **counter-ion**.

gel *n* A semirigid dispersion of a a colloidal solid in a liquid, e.g., a jelly. From a shortened form of *gelatin*.

Gibbs, Josiah Willard (1839-1903) American physical chemist. Founder of chemical thermodynamics. See **Gibbs adsorption equation; Gibbs-Duhem equation; phase rule.**

Gibbs adsorption equation; Gibbs adsorption isotherm The relation between the amount of **surface-active agent** adsorbed at an **interface** with the rate of change of **surface tension** (or **interfacial tension**) with concentration

$$\Gamma = (-1/RT)(\partial\gamma/\partial\ln a_2)$$

where Γ is the **surface concentration** or **surface excess**, γ is the surface or interfacial tension, and a_2 is the activity of the surface-active solute. In dilute systems, the concentration may be substituted for the activity.

Gibbs dividing surface See **dividing surface.**

Gibbs-Duhem equation The relation between the **partial molar quantities** of a system and the concentration of its components:

$$\sum n_i Y_i = 0$$

where n_i and Y_i are the mole fraction and partial molar quantity, respectively, of the *i*th component. Thus, in a two-component system, knowledge of one partial molar quantity enables the calculation of the other. Due to **J. W. Gibbs** and Pierre-Maurice-Marie Duhem (1861-1916), French scientist.

Girifalco-Good equation See **Good-Girifalco equation.**

Good-Girifalco equation For the calculation of interfacial tension from the surface tension of the individual liquids

$$\gamma_{12} = \gamma_1 + \gamma_2 - \Phi(\gamma_{12})^{1/2}$$

where Φ is a function of the molecular volumes of the two liquids, and has an empirical range of 0.55-1.15. Named for **Robert J. Good** and **Louis Anthony Girifalco** (b. 1928). See **Antonoff rule**; **Fowkes equation**.

Good, Robert James (b. 1920) American chemist. See **Good-Girifalco equation**.

gum *n* Any one of a number of viscid and amorphous plant exudates, which harden on exposure to air, usually soluble or miscible with water, forming a viscous liquid. See **protective colloid**.

Gouy-Chapman equation Solution of the **Poisson-Boltzmann** equation for an electric **double layer**, which avoids the restriction to low values of the potential imposed by the **Debye-Hückel** theory. Named for **Louis-Georges Gouy** (1854-1926), French physicist and **David Leonard Chapman** (1869-1958), English physicist.

Gouy-Chapman layer The portion of the diffuse **double layer** existing outside the **Stern layer**.

Graham, Thomas (1805-1869) British chemist. Invented the word **colloid**, by analogy with Gr. *kolla* glue.

Guth-Gold-Simha equation See Appendix F.

H

Hamaker constant   The constant of proportionality in the equation for the attractive potential between colloid particles, arising from **London** or **dispersion forces**. In principle, Hamaker constants can be calculated from fundamental properties of the molecules by means of the **Lifshitz theory**, but in rough calculations a value of the order 10^{-13} ergs works well. At large distances of separation, **retardation effects** must be taken into account. See **Lifshitz-van der Waals forces**.

Hanai equation   See Appendix E.

Harkins, William Draper   (1873-1951).   American physical chemist, noted for his work on, among other things, **monomolecular films** and gas-solid **adsorption**.

Harkins-Jura isotherm   In gas/solid **adsorption,** an equation describing the Type II **isotherm.** For **W. D. Harkins** and G. Jura.

harmonic mean   See Appendix B.

Hatschek equation   For the **viscosity of emulsions** with $\phi > 0.5$

$$\eta = \eta_0[1/(1 - \phi^{1/3}]$$

where η and η_0 are the viscosities of the emulsion and the **continuous phase,** respectively, and ϕ is the **volume fraction.** See **Sibree equation.**

HDC *acronym* *h*ydro*d*ynamic *c*hromatography. Method for the sizing and separation of colloidal particles, e.g., **latices, emulsions,** by forcing the liquid dispersion under pressure ($\approx$20 atm) through a packed column of non-porous beads of radius of $\approx$10 μm. Particles of different sizes travel at different speeds through the bed, and may be collected as fractions.

head group   In a **surfactant,** the group in the molecule attached to the **lipophile** (usually a hydrocarbon chain), which imparts the **hydrophilic** property to the molecule.

heat of adsorption   See **adsorption, heat of.**

HEED *acronym* *h*igh *e*nergy *e*lectron *d*iffraction. Similar to **LEED,** except that electrons in the 18-25 keV are used. The electrons strike the sample at a glancing angle to avoid excessive penetration.

Helmholtz condenser equation The potential across an electrical condenser is given by

$$\Delta V = \sigma d / \epsilon_0 \epsilon$$

where σ is the surface charge on the condenser plate, d is the separation, and ϵ_0 and ϵ are the permittivity of vacuum and the permittivity of the material between the plates (their product being equal to the **dielectric constant**). The Helmholtz condenser may be considered as a first approximation to a **double layer**. See **Helmholtz**; **Helmholtz double layer**.

Helmholtz double layer For small potentials and close to the surface, the **double layer** can be treated as a **Helmholtz condenser**, with a distance of separation equal to the **Debye length**.

Helmholtz, Herman von (1821-1894) German physicist, noted for his investigations into electrostatics and sound.

Helmholtz-Smoluchowksi equation See **electrophoretic mobility**.

hemi-micelle *n* A surface **aggregate**, believed to be formed by **surfactant** molecules adsorbed on a surface above the **cmc**. It is doubtful if such aggregates, if they exist, have properties similar to those of a **micelle**. From Gr. *hemi-*, half + micelle, in reference to their supposed shape.

Henry's equation A correction to the **electrophoretic mobility** which takes into account the size and shape of the migrating particle. For D. C. Henry.

hexagonal close packing The densest packing of uniform spheres, in which the spheres occupy 74% of the volume.

heterocoagulation *n* The act of forming **aggregates** when the particle surfaces have significantly different charge status or when the **aggregates** consist of more than one type of particle.

heterodisperse *adj* Of a system containing many different particle sizes. —**heterodispersity** *n*.

heterogeneous *adj* Composed of diverse elements or constituents, especially of diverse phases, as in heterogeneous **catalysis**. See **homogeneous**.

high energy electron diffraction See **HEED**.

higher order Tyndall spectra See **HOTS**.

Hildebrand, Joel H. (1881-1983) American chemist, especially noted for his work in the theory of solutions. See **cohesive energy density; solubility parameter.**

Hill-de Boer equation For the gas/solid interface, the two dimensional analogue of the **van der Waals** imperfect gas law

$$(\Pi + \alpha/A_2)(A - A_0) = kT$$

in which A_0 is the excluded area, analogous to the van der Waals b, and α the analogue of the constant a.

HLB See **hydrophile-lipophile balance.**

HLB-temperature The temperature at which an **emulsion** (usually stabilized by **nonionic** emulsifiers) undergoes **inversion** from O/W to W/O, or vice versa. The HLB-temperature is related to the **HLB-number** of the emulsifiers used. See **phase inversion temperature.**

Hofmeister, Franz (1850-1922) German scientist.

Hofmeister series See **lyotropic series.**

homogeneous *adj* Composed of elements of the same kind or nature, as in homogeneous **catalysis.** See **heterogeneous.**

homogeneous nucleation temperature In **nucleation,** the temperature at which the probability of forming a nucleus of critical radius by spontaneous fluctuations increases very rapidly with a small decrease in temperature, and hence the rate of the process increases markedly.

homogenizer *n* A device for preparing dispersions in which the liquids (or solid and liquid) are forced through a fine, spring-loaded orifice under pressure. Loosely, any device for this purpose. See **colloid mill; static mixer; ultrasonic homogenization.**

homopolymer *n* A polymer in which the monomeric units are identical, as opposed to a **copolymer** or **block copolymer.**

homotactic *adj* Descriptive of an ideal, crystal-like surface. From Gr. *homos* (similar) + *taxis* (arrangement). Coined by S. Ross.

Hooke, Robert (1652-1703). English scientist.

Hooke's Law   The ideal behavior of an elastic solid under a tensile (stretching) stress γ is given by

$$F/A = Y\gamma$$

where F/A is the applied force per unit area and Y is **Young's modulus**.

HOTS *acronym h*igher *o*rder *T*yndall *s*pectra.   The spectrum observed when a dispersion of spheres is irradiated with white light, different colors being scattered at different angles.

hydration pressure   Repulsion between hydrated **hydrophilic** substances immersed in water; due to the orientation of the water molecules of hydration, which repel one another.

hydrodynamic chromatography   See **HDC**.

hydrophile *n*   A group in a molecule which is preferentially water-soluble, usually through hydrogen-bonding; also a **surface** which is **wetted** by water. —**hydrophilic** *adj.*

hydrophile-lipophile balance; HLB   A method of characterizing **surface-active agents** (especially **emulsifiers**), by assigning a so-called **HLB-number**, which is, in effect, a measure of polarity; thus, surfactants with low HLB numbers are soluble in organic solvents, while those with high HLB numbers are more soluble in water.   The original, rather simple, definition of HLB number, due to W. C. Griffin (b. 1914), was

HLB = Wt% hydrophile/5

The division by 5 has no theoretical significance, but merely reduces the scale to more convenient numbers. Its application to nonionic surfactants is straightforward, but it is less easy to assign HLB numbers to ionic surfactants by the use of this equation. In recent years, it has been suggested that the HLB number is actually a thermodynamic quantity, related to the free energy of micellization, and can thus be calculated from statistical mechanical considerations. See **HLB-temperature; phase inversion temperature, required HLB.**

hydrophilic colloid See **lyophilic colloid.**

hydrophilicity *adj* The property of being **hydrophilic.**

hydrophobic colloid See **lyophobic colloid.**

hydrophobe *n* A group or surface which is not soluble in or wetted by water; not precisely synonymous with **lipophile.** —**hydrophobic** *adj*

hydrophobic bonding The apparently enhanced attraction between two particles or groups in a solvent (e.g., a hydrophobic substance in water), actually arising from the strong solvent-solvent interaction which effectively "rejects" the solute (e.g., the **hydrophobe**).

hydrophobic effect See **hydrophobic bonding.**

hydrophobic interactions See **hydrophobic bonding.**

hydrophobicity *adj* The property of being **hydrophobic.**

hydrotrope *n* A low-molecular-weight **surfactant,** capable of enhancing aqueous solubility, probably acting as a **cosurfactant.** More generally, any substance which enhances solubility, e.g., iodide ions for iodine. From Gk. *hydor* water + *tropos* turning (towards). —**hydrotropic** *adj.* —**hydrotropicity** *n* The property of being a *hydrotrope.* See **microemulsion; solubilization.**

hyperfiltration *n* See **reverse osmosis.**

hysteresis *n* Lag in the response of a body to changes in forces; phenomenon exhibited by a body in which the response to change is dependent on upon its past reactions to change. From Gr. *hystera* womb, used in the sense of inferior (!). See **adsorption hysteresis; contact angle hysteresis.**

I

ideal gas A gas which obeys the Boyle-Charles Law

$$PV = nRT$$

where P is the pressure, V the volume, n the number of moles of gas, T the absolute temperature, and R the gas constant (see Appendix A). Named for Robert Boyle (1627-1691), English chemist, and Jacques A. C. Charles (1746-1823), French physicist.

ideal gas, two-dimensional Of a **monolayer** governed by the two-dimensional form of the **ideal gas** law, i.e.,

$$\Pi A = RT$$

where Π is the **surface pressure** of the monolayer, A is the area of the monolayer expressed as area/mole, R is the gas constant (see Appendix A), and T is the absolute temperature.

IHP *acronym* *i*nner *H*elmholtz *p*lane. In the **Helmholtz double layer**, the plane of charge close to the surface due to adsorbed ions, as opposed to the **OHP (outer Helmholtz plane)**, which is situated where the diffuse layer begins.

image charge When a charge is in proximity to a plane, the effect on the plane may be calculated by substituting for the plane an *image charge*, located at the mirror image position on the other side of the plane. See **image force; image plane.**

image force An attractive force generated by the presence of the **image charge.**

image plane The plane in which an **image charge** is located.

imbibition *n* The uptake of liquid by a **gel** (or porous substance) without change of volume.

immersion, heat of The small amount of heat, measured calorimetrically, evolved when a dry solid is immersed in a liquid. By application of the **Young equation** the heat of immersion can be related to the liquid/vapor **surface tension** γ_{lv} and the liquid/solid **contact angle** θ by the relation

$$-\Delta H_{im} = \gamma_{lv}\cos\theta - T\cos\theta(d\gamma/dT) - T\gamma_{lv}(d\cos\theta/dT)$$

where T is the absolute temperature.

independent surface action   The concept, introduced by I. Langmuir, that, qualitatively at least, it may be supposed that each part of a surface molecule possesses a local **surface free energy** (**surface tension**). See **parachor**.

indifferent electrolyte   In a **colloidal** system, an electrolyte containing no ions in common with those of the colloidal particles. Also **indifferent salt**. See **counter-ion**.

ink bottle pore   A pore in a solid of the shape of an ink bottle, i.e., with a narrow neck opening into a larger volume.

interaction parameter   The quantity in the **Flory-Huggins theory** which is a measure of solute/solvent interaction.

interfacial layer   A nonisotropic interphase between two extended phases, i.e., the layer of **surfactant** which is **adsorbed** at an oil-water interface. See **monolayer; monomolecular layer**.

interfacial tension   See **surface tension**.

interferometry *n*   A measurement based on the fact that the light reflected from the front and back surfaces of a film travel different distances, giving rise to interference effects. Such measurements give information on, e.g., the thickness of the film.

intrinsic viscosity   The limit, at zero concentration, of the **reduced viscosity**.

inverse micelle   A **micelle** in which the **lipophilic** moieties form the micelle surface, as opposed to a *normal* micelle. A nonaqueous micelle.

inversion *n* The process whereby the **continuous and discontinuous** phases of an **emulsion** reverse, e.g., a W/O emulsion becomes O/W. If one assumes, with Wo. Ostwald (1883-1943), that an emulsion is made up of uniform, incompressible spheres, inversion should occur at 74.02% internal phase (closest packing of spheres). The assumption is not, of course, justified, but a surprising number of emulsions will invert near this concentration. See **bicontinuous; phase inversion temperature.**

ion exchange The exchange of ions in solution with those on the surface of a solid, such as a **clay** or synthetic ion exchanger. Ion exchange may be used for purification, or the phenomenon may be used advantageously to prepare or modify a **catalyst.**

Ising lattice A one-dimensional lattice in which the energy depends on the number of *unlike* neighbors.

isobar *n* A mathematical representation of **adsorption** at constant *pressure.* See **isotherm.**

isodisperse *adj* See **monodisperse.**

isoelectric point The pH or salt concentration at which a **colloidal** particle acquires zero charge, i.e., when its **zeta potential** equals zero, and it will no longer move in an electric field. See **electrophoresis, ZPC.**

isoionic point The pH or salt concentration at which a protein has a *net* zero charge, although, strictly speaking, it is not *uncharged.*

isostere *n* The mathematical representation of adsorption at constant *volume*. See **isotherm**.

isotherm *n* A mathematical representation of **adsorption** as a function of concentration (for adsorption from solution) or pressure (for gas/solid adsorption) at a constant temperature (hence the name). For gas/solid adsorption, S. **Brunauer** has identified five types of isotherms, usually designated as Type I, Type II, etc. These isotherms are characterized by their shapes. Type I is the classical **Langmuir isotherm**, corresponding to **monolayer** adsorption. Type II is the characteristic **BET** isotherm. Other than Type I, the isotherms probably correspond to multilayer adsorption.

J

jet impingement A method of emulsification in which two jets of liquid are directed at each other at high velocity. The resulting emulsion usually has a very small average diameter. See **miniemulsion**.

jet, oscillating See oscillating jet. ·

Jones-Ray effect The apparent decrease in **surface tension** of water on the addition of low concentrations of salt, when measured by the **capillary rise method**. **Langmuir** explained this phenomenon as due to the adsorption of a thin layer of water on the surface of the capillary, thus reducing its effective radius, but recent work has cast doubt on this explanation.

85

joule *n* The SI unit of work or energy.

Joule, James Prescott (1818-1899) British scientist.

jump potential See **chi potential**.

K

Kelvin equation The relationship between the vapor pressure of a bulk liquid and that of a small droplet of the liquid

$$RT \ln(P/P^0) = 2\gamma[V]/r$$

where P is the vapor pressure of the bulk, P^0 the vapor pressure of the droplet, γ is the **surface tension**, $[V]$ is the molar volume of the liquid, r is the radius of the droplet, and R and T are the gas constant and the absolute temperature, respectively. See **Laplace equation**.

Kelvin, Lord See **Thomson, William**.

Keesom forces The **interaction energy** between **dipoles**. If the dipoles are freely rotating, the interaction is proportional to the

87

inverse sixth power of the separation. For W. H. Keesom

kink *n* See **defect.**

Krafft temperature See **Krafft point.**

Krafft point The temperature (usually Celsius), or, more precisely, the narrow temperature interval, above which the solubility of a **surfactant** rises sharply. At this temperature the solubility of the surfactant is equal to the **cmc.** Not to be confused with the **cloud point.** Named for F. Krafft, German chemist.

Krigbaum, William Richard (b. 1922). American chemist. See **Flory-Krigbaum theory.**

Kruyt, Hugo Rudolf (1882-1959) Dutch colloid chemist. Editor of the fundamental two-volume *Colloid Science* (1952).

kurtosis *n* The fourth moment of a distribution; a measure of the *asymmetry* of a distribution. See Appendix D.

L

lamella *n* A thin plate, membrane, or film, such as separates the gas cells in a foam. *—pl* **lamellae, lamellas.** From L. *lamella*, dim. of *lamina* a thin plate.

laminar flow Fluid flow in which there is no slip at the surface of shear; non-turbulent flow.

Langevin equation For the motion of a spehrical particle experiencing **Brownian motion**

$$m(\mathrm{d}V/\mathrm{d}t) + 6\pi\eta aV = F(t)$$

where *V* is the velocity, *m* and *a* are the mass and radius, spectively, *η* is the viscosity of the medium, and F(*t*) is the fluctuating Brownian force. This is simply a version of Newton's second law.

Langmuir, Irving (1881-1957) American scientist. In addition to his important work in surface chemistry, Langmuir made major contributions to atomic and molecular theory, catalysis, and more. He was awarded the Nobel prize in 1932.

Langmuir-Blodgett (L-B) film A film formed on a solid surface by the sequential transfer of **monomolecular layers** spread and compressed at a liquid-gas **interface**, accomplished by dipping the solid substrate in and out of the liquid. In addition to the precise control of thickness and composition which the technique affords, the preferred molecular orientation imposed by spreading on the liquid surface may be retained in the L-B film. One use of this technique is in the preparation of coated photographic lenses. See **film balance; Langmuir, Irving; monolayer.**

Langmuir isotherm For gas/solid **adsorption,**

$$p = K(\theta/1 - \theta)$$

where p is the pressure, and θ is the fractional surface coverage. For **I. Langmuir.**

Langmuir trough See **film balance.**

Laplace equation The relation between the pressure differential Δp across a surface and the principal radii of curvature R_1 and R_2

$$\Delta p = \gamma(1/R_1 + 1/R_2)$$

where γ is the **surface tension**. Note that for a sphere $R_1 = R_2$ = R so that $\Delta p = 2\gamma/R$, while for a cylindrical surface $R_1 = \infty$, hence $\Delta p = \gamma/R_2$, and for a plane surface $R_1 = R_2 = \infty$ so that $\Delta p = 0$. Named for **Pierre Simon Laplace**.

Laplace, Pierre Simon, Marquis de (1749-1827) French scientist. Author of *Mécanique céleste*.

latex *n* A (usually) milky dispersion of polymer particles, such as natural or synthetic rubber, or other polymeric materials. —*pl* **latices, latexes.** From L. *latex* water, juice, fluid.

LCFT *acronym* *l*ower *c*ritical *f*locculation *t*emperature. See **CFT.**

LEED *acronym* *l*ow *e*nergy *e*lectron *d*iffraction. The scattering of a low-energy electron beam by a crystal surface. The resulting scattering diagram may be used to determine the crystal structure of the surface. Used in conjunction with **AES,** a crystal surface, e.g., that of a **catalyst,** may be thoroughly characterized.

Lennard-Jones potential Measure of the potential energy between two atoms, given by

$$U(r) = -U_m[(r/r_m)^{-12} - 2(r/r_m)^{-6}]$$

where r is the distance of separation, r_m is the position of the potential minimum, and U_m is the value of $U(r)$ at r_m. This

potential is often referred to as the *6-12 potential*. Named for
Sir John Edward Lennard-Jones (1894-1954), British physicist.

lens *n* Descriptive of the shape of, e.g., an oil droplet on the
surface of water, when the gas/liquid-1/liquid-2 **contact angle**
is finite.

Lewis acid Any substance capable of forming a covalent bond
with a base by accepting a pair of electrons from it. On, e.g.,
a catalyst, this may simply be a lattice site. See **electron acceptor; electron donor; Lewis base.**

Lewis base Any substance capable of forming a covalent bond
with an acid by donating a pair of electrons to it. On, e.g., a
catalyst, this may simply be a lattice site. See **electron acceptor; electron donor; Lewis acid.**

Lewis, Gilbert Newton (1875-1946) American chemist. Noted
for, among other things, the octet theory of covalent bonding,
acid-base theory.

Lifshitz-van der Waals forces Designation for the three collec-
tive electrodynamic forces: orientation (**Keesom**), induction
(**Debye**), and dispersion (**London**) forces. The effects of these
forces was first shown to exist by **J. D. van der Waals** in 1879;
in 1955 **E. M. Lifshitz** demonstrated that, on a macroscopic
scale in the condensed state the three interactions should be
treated in the same manner.

light scattering If a colloidal system is irradiated with a
narrow beam of monochromatic light, the light is scattered at

various angles to the incident beam, owing to the **Rayleigh effect**. Measurement of the intensity of the scattered light as a function of angle permits the calculation of the size (and, under certain conditions, the shape) of the scattering particle. Widely used for determining the molecular weight of polymers and, e.g., **micelles**. See **PCS**.

limiting viscosity number IUPAC designation for **intrinsic viscosity**.

linear mean diamter See Appendix B.

Lineweaver-Burke method A graphical method, used to analyze data from **micellar catalysis**.

lipid *n* Any one of a group of organic compounds, character-ized by a greasy feel, insoluble in water, but soluble in a number of organic solvents, such as alcohol or ether; a fat. From Gk. *lipos* fat.

lipid bilayer See **phospholipid bilayer**.

liquid crystal A liquid phase which possesses some degree of molecular order, as evidenced by **birefringence**. Frequently, encountered at the **interface** in **microemulsions**.

lipophile *n* A group in a molecule which is preferentially soluble in organic solvents (and hence insoluble in water); also a solid which is **wetted** by organic liquids. —**lipophilic** *adj.*

lipophilicity *n* The property of being **lipophilic**.

lipophobe *n* A group in a molecule which is *insoluble* in organic solvents; also a solid which is *not* **wetted** by organic solvents (but not necessarily wetted by water). —**lipophobic** *adj*. See **hydrophile.**

lipophobicity *n* The property of being **lipophobic.**

Lippmann equation The basic equation of **electrocapillarity**

$$(d\gamma/dE)_{\mu,P,T} = -\sigma_0$$

where γ is the **surface tension**, E is potential, μ is the chemical potential of the dissolved salt, and σ_0 is the charge per unit area. See **electrocapillary maximum.**

logarithmic viscosity number IUPAC designation for **inherent viscosity.**

London forces The **interaction energy** between two induced dipoles, proportional to the inverse sixth power of the distance. If the dipoles are separated by a distance such that **retardation** effects are called into play, the energy is proportional to the inverse seventh power.

loops, polymer In an adsorbed polymer, polymer segments attached to the surface at two points, forming a loop extending into the solvent. See **tails, polymer; trains, polymer.**

low energy electron diffraction See **LEED.**

lubrication *n* The application of an oily or greasy substance to an **interface** in order to reduce **friction.** If the layer of lubricant is thick enough, it acts, e.g., to substitute liquid/solid contact for solid/solid contact, thus reducing the coefficient of friction. For a very thin layer of lubricant, *boundary layer lubrication* takes place. —**to lubricate** *v.*

Lundelius rule The observation that there is an inverse relation between the solubility of a solute and the extent of its **adsorption.**

lyophilic colloid An almost obsolete designation for a *single-phase* colloid, e.g., a polymer solution, as contrasted with a **lyophobic colloid.** From Gr. *lyein* to release, to dissolve + *philos* to love; hence, *solvent-loving.* See **hydrophilic.**

lyophobic colloid An almost obsolete designation for a *two-phase colloid,* e.g., a sol, as contrasted with a **lyophilic colloid.** From Gr. *lyein* to release, to dissolve + *phobod* to hate, to fear; hence, *solvent-hating.* See **hydrophobic.**

lyophobic mesomorphic phases See **liquid crystal.**

lyoschizophrenia *n* Descriptive of the behavior of a **surface-active agent** in solution, i.e., not knowing, as it were, in which kind of solvent it is soluble. From Gk. *lysein* to dissolve + *schizo* I cut + *phren* understanding. Coined by P. Becher.

lyotropic series A measure of the effect of various ionic species on colloidal systems, e.g., coagulation of gels, lowering of CMC, etc. Although there are minor variations, the series, expressed in descending order of effectiveness is quite reproducible. Typically, for anions

$$CNS^->I^->Br^->Cl^->F^->NO_3^->ClO_4^-$$

while for cations it is

$$Cs^+>Rb^+>K^+>Na^+>Li^+$$

The effect is evidently connected with the size and polarizability of ions. Also called the **Hofmeister series**. See **Hofmeister, Franz; Schulze-Hardy rule.**

M

macroemulsion See **emulsion**.

macroion *n* A colloidal particle possessing a charge attributable to the presence of ionic groups.

macromolecule *n* A synthetic or natural high polymer. The molecular weight at which the designation of a molecule changes from simply *large* to *macromolecule* is not well-defined. It is possible (and sometimes useful) to think of a solution of macromolecules as a **colloidal** system. —**macromolecular** *adj*. See **oligomer**.

macropore *n* In a solid, pores with a width greater than about 100 nm. See **mesopore**; **micropore**.

Marangoni, Carlo (1840-1925) Italian scientist.

Marangoni effect The restoring force generated in a **film** or **monolayer** on expansion, arising from the increase in **surface tension** brought about by the lowered surface concentration. Named for **C. Marangoni**. See **tears of wine**.

Mark-Houwink coefficient See **viscosity average molecular weight**.

Martin's diameter For irregularly-shaped particles, the length of the line which bisects the projected area of the particle. The direction in which the line is drawn is arbitrary, but should be used consistently.

mass area mean diameter A mean in which the property averaged is the particle area. See Appendix B.

maximum bubble pressure A method of measuring **surface tension** in which a bubble is formed in the liquid by blowing an inert gas through a small tube. When the shape of the bubble is exactly hemispherical, i.e, is equal to that of the tube, the pressure differential ΔP has its maximum value. Since the tube may be placed some arbitrary distance t below the surface of the liquid,

$$\Delta P_{max} = P_{max} - P_t$$

where P_{max} is the measured pressure, P_t is the pressure corresponding to the hydrostatic head. Then if ΔP_{max} is represented in terms of the height of a corresponding column of liquid, i.e.,

$$\Delta P = \Delta \rho g h = 2\pi r \gamma \cos \vartheta$$

as in the case of **capillary rise**. Exact determinations require the use of the **Bashforth** and **Adams** correction terms. As a practical matter, this is not a convenient technique.

McBain, James William (1882-1953) British-American chemist.

mean *n* An average, especially the **arithmetic mean**. See Appendix B.

mean curvature For a curved surface, the mean curvature *J* is given by

$$J = 1/R_1 + 1/R_2$$

where R_1 and R_2 are the **radii of curvature**.

mean, geometric See Appendix B.

mean, harmonic See Appendix B.

mean ionic activity Defined as

$$a_\pm{}^v = a_+{}^{v+} a_-{}^{v-}$$

where a_+ and a_- are, respectively, the activities of the positive and negative ions, and $v = v_+ + v_-$, i.e., the number of ions present when the molecule is completely dissociated.

membrane *n* A thin, pliable film, which may be of animal or vegetable origin, or a synthetic material, e.g., a polymer film.

membrane, ion-selective In electrodialysis, a membrane rendered impermeable to ions of a specific charge by imposing a charge of the same sign on the membrane, e.g., a positive charge for anions.

membrane potential See **Donnan effect**.

membrane, semipermeable See **semipermeable membrane**.

meniscus *n* The convex or concave upper surface of a column of liquid, the curvature of which is caused by **surface tension**; the shape of a liquid surface in contact with a solid. —**meniscoid** *adj*.

mesomorphic phase A micellar structure which may be described as possessing one or two infinite dimensions. See **liquid crystal**.

mesopore *n* In a solid, a pore with a width intermediate between those of a **micropore** and a **macropore**.

metastable *adj* Descriptive of a system existing at an energy level (i.e., in a *metastable state*) above that of a more stable state and requiring only the addition of a small amount of energy to enable transition to the more stable state. —**metastability** *n*.

micellar catalysis See **catalysis, micellar**.

micellar emulsion See **microemulsion**.

micellar solution A solution containing **micelles**. See **critical**

micelle concentration.

micelle *n*   An oriented aggregate of surface-active molecules formed in solution when the solubility limit for single molecules (monomers) has been reached. Such aggregates may contain from approximately 10 to approximately 100 monomer molecules which are oriented in the sense that either the **hydrophile** or **lipophile** moieties are on the outside of the micelle, depending on the solvent. Thus, in water, the hydrophilic portion of the surfactant molecules are on the outside, and such a micelle is usually termed a normal micelle; in an organic liquid, the lipophilic moieties are on the outside, and the micelle is termed an **inverse** or **inverted micelle**; these distinctions are useful, but semantically rather odd, since there is nothing especially normal about a *normal micelle*. The shapes of micelles may vary; spheres, discs, and rods are all possible. The particular shape which the micelle adopts depends on a variety of factors, temperature, concentration, and, quite importantly, the geometry of the surfactant monomer. See **critical micelle concentration**.

micelle, mixed   Micelle composed of more than one type of **surfactant** molecule.

micelle, swollen   A micelle which has been increased in size, either by the presence of **solubilized** molecules or by the formation of a **mixed micelle**. It has been proposed that a **microemulsion** droplet is really a swollen micelle.

micellization *n*   The process of formation of **micelles**.

microelectrophoresis *n*   The direct viewing of **electrophoresis** by means of microscopic observation.

microemulsion *n*   A thermodynamically stable transparent solution of micelles swollen with solubilizate.   Microemulsions usually require the presence of both a surfactant and cosurfactant, the latter of which is usually a fairly short-chain alcohol, and are characterized by the presence of extremely low **interfacial tensions.**   Properly speaking, a microemulsion is not an **emulsion** in the ordinary sense, and, in addition, should be distinguished from **miniemulsions.**

microencapsulation   Process by which a small particle or, more usually, a small droplet is enclosed in a rigid or semi-rigid coating.   The original process involved using a protein, e.g., gelatin, for the encapsulating agent, and depositing it at a liquid/liquid interface by **coacervation.**

micropore *n*   Very small pores (100 - 150 nm in diameter) found in certain types of **adsorbents**, e.g., silica gel, carbons, **zeolites.**

microscope *n*   A device for magnifying small objects.   The simplest type of microscope is the optical microscope, using glass lenses, the invention of which is ascribed to Anthony van Leeuwenhoek (1632-1723), Dutch scientist. See **AES; LEED; SEM; TEM.**

microscope, electron   See **SEM; TEM.**

microscopy *n*   Examination of a material by means of a **microscope.**

microtome method A method of measurement of the **surface concentration** of a liquid by actually slicing a layer from the surface with a knife mounted on a carriage, which could be moved with great speed. Subsequent chemical analysis of the skimmed layer permits a calculation of the surface concentration. Developed by **J. W. McBain**.

middle phase A form of **liquid crystal**, in which the **surfactant** molecules are arranged as cylindrical rods. See **neat phase**.

Mie scattering In **light scattering**, the behavior observed when the scattering particles become comparable in size to the wavelength of the incident light. After Gustav Mie (1868-1957).

mill, colloid See **colloid mill**.

Miller indices A set of three integers defining the orientation and position of a crystal face in terms of the reciprocals (reduced to lowest terms) of the intercepts of the face with each of the crystal axes, e.g., the face of a cubic crystal would have the Miller index 100. After William Hallowes Miller (1801-1880), British scientist.

miniemulsion *n* An emulsion, with droplet sizes in the range of 100 to 1000 nm, reportedly thermodynamically stable. To be distinguished from **microemulsions**.

mixed micelle See **micelle, mixed**.

mobility, electrophoretic See **electrophoretic mobility**.

modulus *n* A coefficient corresponding to a particular property, as, e.g., **modulus of elasticity.**

modulus of elasticity The ratio between an applied stress (or force per unit area) and the resulting fractional deformation, i.e., the ratio of stress to strain.

modulus, shear See **shear modulus.**

modulus, Young's See **Hooke's law.**

molecular sieve See **zeolites.**

moments of a distribution See Appendix B.

monolayer *n* The term usually reserved for a monomolecular film, i.e., an adsorbate film one molecule thick, spread on a liquid substrate. With the proper apparatus, the monolayer may be compressed and expanded and its properties measured. See **Langmuir trough; monomolecular layer.**

monolayer, saturated A monolayer of maximum surface concentration.

monomolecular film; monomolecular layer A **monolayer.** Films that form at surfaces or interfaces are of special importance; such films may reduce friction, wear, and corrosion, or may stabilize emulsions, foams, or solid dispersions. Monolayers on water surfaces reduce evaporation losses. The broad field of **catalysis** involves chemical reactions that are accelerated in the monolayers of reactants at interfaces. Thin films containing proteins, cholesterol, etc. constitute biological membranes. See **cell membranes; Langmuir trough.**

monopolar *adj* Characteristic of a **polar** substance or a surface which has either only **electron-acceptor** and no (or virtually no) **electron-donor** properties, or vice versa. The **surface tension** of a monopolar substance is identical to (or very close to) the **Lifshitz-van der Waals** component of its surface tension.

Monte Carlo method A technique for the numerical approximation to the solution of a mathematical problem by studying the distribution of a randomly-generated variable, usually through the use of a computer.

Mooney Equation For the viscosity of an emulsion

$$\ln \eta/\eta_0 = 2.5\phi/(1 - z\phi)$$

where η and η_0 are the viscosities of the emulsion and of the continuous phase, respectively, ϕ is the **volume fraction**, and z is the so-called *self-crowding factor*.

Mossotti-Clausius equation See **Clausius-Mossotti equation**.

moving boundary electrophoresis A method of determining electrophoretic mobility based on the observation of the motion of a liquid boundary by means of schlieren optics. In the case of a mixture of colloidal particles, e.g., proteins, the individual components can be observed as peaks. See **Tiselius, A.**

multilayer adsorption See **adsorption; isotherm**.

N

Navier, Claude-Louis-Marie-Henri (1785-1836). French scientist.

Navier-Stokes equation   The fundamental equation of fluid flow. In vector form

$$\rho(\partial v/\partial t + v\nabla v) = F - \nabla p + \eta\nabla v$$

where ρ is the density, v is the velocity vector, F is any long-range force acting on the liquid, and p is the pressure (∇p being the pressure gradient). In ordinary flow, F is simply the effect of gravity ρg, where g is the acceleration of gravity. Named for C.-L.-M.-H. Navier and G. G. Stokes.

neat phase A form of **liquid crystal**, in which the **surfactant** molecules are arranged as lamellar micelles. See **middle phase**.

negative adsorption In **DLVO** theory, it is expected that, for example, in the region near a negatively-charged surface the concentration of negative ions should decrease, and hence the bulk concentration of negative ions should increase by the same amount. Measurement of this effect leads to a simple method for the determination of the **surface area**. Also, the case in which the **surface excess** has a negative value.

Newton, Sir Isaac (1642-1727) British scientist, formulator of the laws of gravitation.

Newtonian flow Liquid flow in which the **coefficient of viscosity** is given by **Newton's law**.

Newton's law Describes the flow of fluids under simple shear, i.e.,

$$F/A = \eta(dv/dy)$$

where F is the shearing force applied to an area A, dv/dy is the flow gradient, and η is the coefficient of viscosity (or, briefly, the **viscosity** of the liquid). See **non-Newtonian flow; viscosity**.

non-draining *adj* Term used to describe a polymer coil whose interior is unperturbed by flow, i.e., behaves a like a rigid body whose volume is proportional to r_g^3, where r_g is the **radius of gyration** of the coil.

nonionic *adj* Describing a **surface-active agent** containing no ionic groups. Typically, a nonionic surfactant contains a lipophilic moiety with an active hydrogen which reacts with a plurality of ethylene oxide molecules, thus supplying the hydrophilic portion of the molecule. Thus, typically, $C_{12}H_{25}(OC_2H_2)_nOH$, where n is a number representing the stoichiometry of the preparation. Normally, n is an average, and the surfactant consists of a distribution of ethylene oxide chain lengths, expressed approximately as a **Poisson distribution**. Certain manufacturing processes permit narrower distributions, and careful laboratory syntheses may produce essentially monodisperse materials. —**nonionic** *n* Ellipsis for *nonionic surfactant*. See **anionic; cationic**.

Non-Newtonian flow Liquid flow in which **Newton's law** is not obeyed. See **Newtonian flow; viscosity**.

non-wetting *adj* Used to describe a **surface** for which the **work of cohesion** is greater than the **work of adhesion** of the wetting liquid; thus, the liquid forms a finite **contact angle**.

normal distribution See Appendix D.

nucleation *n* The growth of particles from solution owing to the presence of *nuclei*, i.e., extremely small particles, on to which the nucleating material is adsorbed. If the nuclei are chemically identical with the molecules adsorbed, the nucleation is said to be **homogeneous**; when, however, nucleation is induced by foreign particles, e.g., dust, it is said to be **heterogeneous**. —**nucleate** *v*.

number area mean diameter See Appendix B.

number average molecular weight See Appendix C.

number length mean diameter See Appendix B.

number volume mean diameter See Appendix B.

numerical aperture A measure of the light-gathering power of a lens, given by

$$N.A. = n \sin(A.A.)/2$$

where n is the refractive index of the medium, and $A.A.$ is the *angular aperture* (the angle between the most divergent rays which can pass through the lens to form the image). See **resolving power**.

O

oblate *adj* Flattened at the poles, as of spheroid generated by the rotation of an ellipse about its shorter axis. See **ellipsoid; prolate.**

oblate spheroid See **ellipsoid; oblate.**

Oden balance A device for measuring the **cumulative distribution** curve of a **dispersion** by measuring the total mass of sediment accumulating on a balance pan suspended in the dispersion.

Ohnesorge equation If a liquid 1 is forced through an orifice as a jet into a second liquid 2, e.g., oil into water, there is a critical jet velocity which must be exceeded for breakup of the jet of liquid 1 into droplets defined by

$$\eta_1/(\rho_1\gamma_1 D)^{1/2} = 2000(\eta_1/V_0\rho_1 D)^{4/3}$$

where V_0 is the critical velocity, D is the orifice diameter, and η_1, ρ_1, and γ_1 are the viscosity, density, and surface tension, respectively, of liquid 1.

OHP *acronym outer Helmhotz plane.* In the **Helmholtz double layer**, the plane at which the diffuse layer begins. See **IHP**.

oil recovery Typically, 30-50 per cent of the petroleum in an oil formation cannot be recovered by conventional means. Much research has been applied to the use of **surfactants**, usually as **microemulsions**, to the extraction of this residual material. It is usually referred to as **EOR**, or as *tertiary oil recovery*. It must be admitted, however, that, except in special cases, economic extraction has not been completely successful.

oligomer *n* A **macromolecule** in which the degree of polymerization is approximately <10; a small polymer.

Oliver-Ward equation See Appendix F.

orientation force See **Keesom forces**.

oriented wedge The concept that a **surfactant** molecule adsorbed at an oil/water interface in an emulsion would have its polar end in the aqueous phase. As a consequence, the surface molecules could be considered to be wedge- or cone-shaped, and would therefore stabilize the type of emulsion in which the larger end of the molecule extends outward. Although the concept is consistent with at least some emulsion systems, it is obviously too simplistic to be of value. Due to **W. L. Harkins.** See **Bancroft rule; HLB.**

orthokinetic *adj* Of **aggregation** or **flocculation.** If the collisions leading to aggregation or flocculation are due to external forces, e.g., sedimentation, convection, stirring, etc., the aggregation is described as orthokinetic. See **perikinetic.**

oscillating shear **Shear** arising from the application of a periodic **strain** of frequency w (rad sec^{-1})

$$\gamma = \gamma^{\circ} \sin wt$$

where γ° is the maximum amplitude of the strain. See **Weissenberg rheogoniometer.**

osmometer *n* A device for the measurement of **osmotic pressure.** See **osmometry.**

osmometry *n* The measurement of **osmotic pressure.** See **osmosis; semipermeable membrane.**

osmosis *n* The flow of a fluid through a **semipermeable membrane** from a solution of lower solute concentration to one of higher solute concentration, the effect being to equalize the concentrations. From Gk. *osmos* push, thrust.

osmosis, reverse See **ultrafiltration.**

osmotic pressure The pressure required to raise the vapor pressure of a solvent in a solution to that of the pure solvent.

If a solution and the pure solvent are separated by a **semipermeable membrane**, i.e., permeable by the solvent but not the solute, the osmotic pressure is the hydrostatic head required to prevent **osmosis**. The osmotic pressure is defined by the **van't Hoff** equation

$$\Pi = m_s RT$$

where Π is the osmotic pressure, m_s is the solution molality, and R and T have their usual meanings. See **osmometry**.

osmotic pressure, two-dimensional Representation of the compression behavior of a **monolayer** by the assumption that the **surface pressure** π is equivalent to a three-dimensional pressure P, i.e.,

$$P = \pi/\tau$$

where τ is the thickness of the surface layer; of the order of 1.0 nm for monolayers.

Ostwald ripening Process whereby large particles in a suspension grow at the expense of the smaller ones, owing to the greater solubility of the smaller particles. Named for W. Ostwald. See **Kelvin equation**.

Ostwald viscometer See **capillary viscometer**

Ostwald, Wilhelm (1853-1932) German physicist, Nobel Prize 1909.

outer potential See **Volta potential**.

outgassing *n* The removal of the last traces of adsorbed gases in an **ultra-high vacuum** system, by prolonged baking (10-20 hr) at 300-500°C.

overlap *n* In **DLVO theory**, the interaction of two **double layers** as they approach one another.

overvoltage *n* The voltage excess over the **reversible potential** frequently required to pass appreciable current through an electrochemical cell. See **electrochemical potential**.

P

---

pairwise summation Procedure employed in evaluating the interaction between particles, in which it is assumed that the total interaction is made up of the interactions between pairs of particles, i.e.,

$$V^{1..N} = 1/2 \sum_{\substack{i=1}}^{N} \sum_{\substack{j=1 \\ \neq i}}^{N} V^{ij}(R_{ij})$$

where $V^{1..N}$ is the total interaction energy and $V^{ij}(R_{ij})$ is the interaction energy of molecules i and j separated by the distance R_{ij} in the *absence* of any other molecules.

palisade layer In an aqueous **micelle**, the hydrated shell between the inner core and the polar heads of the **surfactant** molecules constituting the micelle; so-called because of a fancied resemblance to a palisade fence.

parabolic flow The flow profile in a capillary under **Poiseuille flow**.

parachor *n* A molecular quantity, proposed by S. Sugden (1929), defined as

$$P = M\gamma^{1/4}/\Delta\rho$$

where M is the molecular weight, γ is the surface tension, and $\Delta\rho$ is the liquid density minus the vapor density. The parachor may be regarded as a molar volume corrected for the compressive effects of intermolecular forces. It was supposed that the parachor could be represented as a sum of *group* or *atomic* parachors, and thus be useful as a tool in determining molecular structures, e.g., in deciding between isomeric structures. Unfortunately, the additivity of the group parachors is only approximate.

paramagnetic *adj* Of a substance which, placed in a magnetic field exhibits magnetization directly proportional to the applied magnetic field; a substance in which the magnetic moments are not aligned. See **ferromagnetic**.

partial molar quantity A thermodynamic property, defined as

$$Y_i = (\partial Y/\partial n_i)_{n_j,T,P}$$

where Y is the property (e.g., Gibbs free energy, enthalpy, volume, etc.), n_i is the mole fraction of component i, and n_j is the concentration of the other components, i.e., $n_j \neq n_i$. See **Gibbs-Duhem equation.**

partial wetting See wet.

particle size distribution A curve or function describing the distribution of particle (or droplet) sizes around a mean in a dispersion or emulsion. See Appendix D.

partition function In statistical mechanics, the statistical summation of all of the states of a system

$$Q = \sum_i g_i \exp[-\epsilon_i/kT]$$

where g_i is the statistical weight of the ith state and ϵ_i is the energy associated with that state. It may be shown that the Helmholtz free energy is connected with the partition function by the simple relation

$$A = -RT \ln Q$$

passivation n The act of treating a metal surface so as to reduce or inhibit **corrosion.** Passivation may occur spontaneously, as in the passivating oxide film on aluminum. — **passivate** v.

PCS *acronym photon correlation spectroscopy.* A form of light scattering, in which advantage is taken of the fact that when a photon hits a scattering particle the reradiated light has a slightly different frequency from that of the incident light, owing to the Doppler effect arising from the (**Brownian**) motion of the particle. Thus quasi-elastic scattering can be used to determine the **diffusion coefficient** of the scatterer, and, hence (through the application of the **Stokes-Einstein equation**), its size. Also referred to as **quasi-elastic light scattering (QELS)**.

Peclet number A dimensionless quantity corresponding to the ratio between mass transfer and diffusive transfer in flow. If the number is small, **Brownian motion** predominates.

pearlescence *n* The property of having an iridescent luster resembling that of a pearl. —**pearlescent** *adj*. A better etymology would have yielded *nacrescence*.

pendant drop method Method for the measurement of **surface tension**, by analysis of the shape of a drop hanging from the tip of a capillary.

penetration *n* The process whereby a more-or-less surface-active molecule present in the substrate enters into a spread **monolayer**, thereby changing the properties of the monolayer.

peptize *v* To **deflocculate**, to reverse **flocculation**, to **disperse**.

percolation *n* The process whereby a transport property, e.g., conductivity, in a disperse system increases sharply at some concentration, owing to the presence of continuous conducting paths (so-called *infinite clusters*). See **percolation threshold**.

percolation threshold The phase concentration at which **percolation** is observed to occur.

perikinetic *adj* Of **aggregation** or **flocculation**. If the collisions leading to aggregation or flocculation are due solely to **Brownian motion**, the aggregation is described as perikinetic. See **orthokinetic**.

permeability *n* See **Darcy's law.**

permittivity *n* The quantity ϵ in the equation for the force acting between two point charges in a dielectric medium

$$F = Q_1 Q_2 / \epsilon r^2$$

where Q_1 and Q_2 are the charges, and r is the distance of separation. The **dielectric constant** of a substance is the product of its permeability with that of vacuum, i.e., $\epsilon \epsilon_0$. See Appendix A.

persorption *n* A term due to **McBain**, describing the situation in which the pores in a porous **adsorbent** are sufficiently small to act as **molecular sieves**, and hence different apparent **surface areas** would be obtained with **adsorbates** of differing molecular size. Also, in general, the deep penetration of a liquid into a porous solid.

PES *acronym* photoelectron spectroscopy. See **ESCA.**

phase *n* A continuous region of matter, bounded by an **interface**, homogeneous as to composition and thermodynamic properties.

phase diagram A map of the **phase** behavior of a system. For a two-**component** system such a diagram may show behavior as a function of composition versus temperature; for a three-component system, a triangular diagram (due to **J. W. Gibbs**) may be used. This diagram shows the behavior at a single temperature, and several such diagrams may be required to completely describe a system. See **pseudo-phase diagram**.

phase map See **phase diagram**.

phase rule A rule governing the relationships in a multiphase system. Defined as

$$F = C - P + 2$$

where F is the number of **degrees of freedom**, C is the number of **components**, and P is the number of **phases**. Due to **J. W. Gibbs**.

phase-transfer catalysis In the case of an organic reaction in which one reactant is water-soluble and the other soluble in an organic solvent, reaction is facilitated by the use of a phase-transfer **catalyst** (usually a **cationic surfactant**), which enables transfer of the water-soluble reactant into the organic phase. Not be confused with **micellar catalysis**, since it is possible that **micelles** play no role in the process; rather the interphase transport is due to the formation of an ion-pair.

phase volume In a multiphase system, e.g., an emulsion, the volume fraction of the components, such that the sum of the phase volumes is equal to unity. The phase volume is sometimes represented as a volume per cent, in which case, of course, the sum of the phase volumes is 100.

phospholipid *n* Any of a group of fatty compounds, e.g., lecithin, composed of lipid phosphate esters. Used as **emulsifiers**.

phospholipid bilayer A two-layer arrangement of **phospholipid** and **lipid** molecules, with the **hydrophilic** phosphate groups facing outward and the **hydrophobic** lipids facing inward. See **cell membrane**.

photoelectron *n* An electron emitted from a surface on irradiation.

photon correlation spectroscopy See **PCS**.

physical adsorption Adsorption characterized by **adsorbate-substrate** interactions arising from **van der Waals-Lifshitz** forces. See **chemisorption**.

physisorption See **physical adsorption**.

plait point The point in a three-component phase diagram at which a pair of conjugate solutions on a binodal (two-phase region) curve have the same composition. Sometimes incorrectly referred to as *critical point*, which, properly speaking, is a temperature.

Planck's constant The fundamental constant of quantum mechanics, the ratio of the energy of one quantum of radiation to the frequency of the radiation. See Appendix A.

Planck, Max Karl Ernest (1858-1947) German physicist. Awarded Nobel Prize 1918.

plastic flow Deformation or creep of a solid under stress, in which the solid exhibits some of the characteristics of a plastic or liquid.

chemist, noted for his investigations of thin films.

Plateau's problem   The calculation of the shape of a **meniscus** when there is no pressure across the interface, e.g., when the **radii of curvature** meet the condition $1/R_1 = -1/R_2$, as in the case of an open soap film.

PLAWM *acronym*   Pockels-*L*angmuir-*W*ilson-*M*cBain.   See **Langmuir trough**.

Pockels, Agnes (1862-1935)   German schoolteacher, credited with the invention of the **film balance**.

Pockels point   A **surface area** of about 20 Å^2 per molecule, corresponding to the area per molecule in a compressed monolayer of fatty acid, and first reported by **Pockels**.

Pohlman whistle   A device for **ultrasonic emulsification**, in which ultrasonic vibrations are set up by the rapid flow of a jet of the liquids to be emulsified past a thin blade, supported at two points separated by half the wavelength of the desired frequency.   The flow of liquid induces the ultrasonic oscillation in the blade.

point of zero charge   See **zero point of charge**.

Poiseuille equation   For the rate of flow of liquids in a capillary, where the volume rate of flow is given by

$$V/t = (\rho g l + \Delta p)\pi R^4/8\eta l$$

where V is the volume of liquid, t is the time, ρ is the liquid

$$V/t = (\rho g l + \Delta p)\pi R^4/8\eta l$$

where V is the volume of liquid, t is the time, ρ is the liquid density, g is the acceleration of gravity, l is the length of the capillary, Δp is the hydrostatic head driving the liquid, R is the radius of the capillary, and η is the viscosity. Since, with the exception of the density and viscosity, all the terms on the right are characteristic of the capillary, by using the same volume of liquid in a fixed capillary, the equation may be written

$$\eta = A \rho t$$

where A is a constant characteristic of the capillary, and may be calculated from the dimensions of the capillary, or determined empirically by the use of liquids of known density and viscosity. For exact work, end corrections must be applied. This is the basis of the **Ostwald viscometer.**

Poiseuille flow Flow in which the flow pattern takes on a parabolic form, with flow slowest at the walls of the capillary and most rapid in the center.

Poiseuille, Jean-Louis-Marie (1799-1869) French physician, noted for his studies of the flow of blood.

Poisson distribution A probability distribution,[*] due to the French mathematician S. D. Poisson (1781-1840), expressed

---

[*] Not to be confused with the Poisson equation of electrostatics, for which see **DLVO theory.**

mathematically as

$$N_x = e^{-n}n^{x-1}/(x - 1)!$$

where, for the case of the formation of ethylene oxide-containing surfactants, N_x equals the number of molecules of surfactant containing x ethylene oxide (EO) units and n is the total stoichiometric amount of EO involved in the reaction. See **nonionic**. See also Appendix D.

Poisson equation; Poisson-Boltzmann equation n See **double layer**.

polar *adj* Characteristic of a substance whose molecules posses a permanent **dipole moment**. Also, of a substance or surface whose functionality is described as being **electron-acceptor** or **electron-donor**, or both. The use of *polar* in this latter sense may lead to confusion, and should be discouraged.

Polanyi adsorption isotherm A model for multilayer **adsorption**, which assumes that there is **potential** field at the surface of a solid, into which the **adsorbate** molecules "fall." Named for Michael Polanyi (1891-1976).

polarization n In physical chemistry,[*] a vector quantity indicating the electric **dipole moment** per unit volume of a **dielectric**. —**polarizability** *adj*.

polarization, adsorption In **chemisorption**, the possibility that

---

[*] Unfortunately, polarization is used in science with a number of different connotations. Caveat lector!

adsorption bond formation may **polarize** the **adsorbate** or strongly orient an existing **dipole**, leading to electrostatic repulsion in the adsorbed layer.

polarization, electrode See **overvoltage.**

polarize *v* To bring about **polarization.**

polydisperse *adj* Having a distribution of sizes. See **monodisperse.**

polyederschaum *n* A concentrated foam, in which the bubbles are in the form of polyhedra, characterized by **Plateau borders.** From Ger. *polyeder* polyhedron + *schaum* foam. See **foam; kugelschaum.**

polyelectrolyte An electrolyte of high molecular weight, usually polymeric, either of natural or synthetic origin.

polymerization, degree of The average number of monomer units making up a polymer chain. Frequently abbreviated DP.

polywater See **water, anomalous.**

pore size distribution The distribution of pore radii in a porous solid.

porosimeter *n* A device for the determination of **pore size distribution,** usually by the measurement of pore penetration by mercury under pressure. —**porosimetry** *n*.

potential barrier; potential maximum A maximum in a **poten-**

tial energy diagram, which must be exceeded in order to pass from one region to another. See **activation energy; potential minimum**.

potential-determining ions At an interface, those species of ions which, by virtue of their equilibrium distribution between the two **phases** (or by their equilibrium with electrons in one of the phases) determine the difference in **Galvani potential** between the two phases.

potential, electric See **electric potential**.

potential energy diagram A plot of the potential energy of a system against, e.g, distance. In **DLVO theory**, for example, the effect of distance on the interaction of **colloidal** particles is shown by means of a potential diagram. See **activation energy; potential barrier**.

potential minimum A minimum in a **potential energy diagram**. If the potential energy at that point is negative, it corresponds to a metastable state of the system.

primary minimum The minimum of the potential energy diagram in **DLVO theory** corresponding to zero distance of separation, e.g., coalescence of the colloidal particles. See **secondary minimum**.

principal radii of curvature See **radius of curvature**.

probe molecule A substance, solubilized in a **micelle**, whose optical absorption or fluorescence may be measured spectroscopically to give information about the micelle, e.g., the **CMC**.

projected area diameter The area of a circle having the same area as an irregularly-shaped particle, viewed normally to a plane surface on which the particle is at rest in a stable position. See Appendix B.

prolate *adj* Elongated along the polar diameter, as a spheroid generated by revolution of an ellipse about its longer axis. See **ellipsoid; oblate.**

promoter *n* Constituent of a **catalyst** surface which has the property of shifting the equilibrium or lowering the activation energy, so as to *promote* the production of the desired product.

protect *v* To make an **electrocratic** system less susceptible to electrolyte by adsorption of a **protective colloid.**

protective colloid A **hydrophilic** polymer, such as a **gum** or gelatin, used to **protect** a **colloid** system.

PSD *acronym* *p*article *s*ize *d*istribution. See **particle size distribution.**

pseudo-phase diagram A triangular **phase diagram** used for a system containing more than three **components.** In such a diagram, two or more components are maintained at a constant composition ratio, and this ratio is treated as a single component.

pseudoplasticity *n* The non-time-dependent decrease in apparent **viscosity** with increasing **rate of shear,** in the absence of a **yield value.** See **thixotropy.** —**pseudoplastic** *n* A substance

Q

QELS *acronym* *q*uasi-*e*lastic *l*ight *s*cattering. See **PCS**.

quadrupole *n* In the simplest (linear) case, a quadrupole consists of a charge of magnitude $+2e$ (where e is the electronic charge), with two charges of magnitude $-e$ at distances $\pm a$ from the origin. Two types of quadrupole interactions are possible; quadrupole-dipole and quadrupole-quadrupole. Since quadrupole interactions fall off approximately as the inverse ninth power of the distance, their effect is minimal. Introduced by **P. Debye**. See **Lifshitz-van der Waals forces**.

R

---

radius of gyration The radial distance from an axis at which the mass of the body can be assumed to be concentrated, and at which the moment of inertia is equal to that of the body, defined as

$$R_g^2 = \sum_i m_i r_i^2 / \sum_i m_i$$

random coil A long-chain molecule oriented in a three-dimensional **random walk**. Such coils contain solvent molecules in their interior.

random walk The motion of an object in which steps of equal length are taken, but where the direction of each step is random; sometimes referred to as the *drunkard's walk*, for obvious reasons. Although the distance from the starting point after n steps cannot be calculated, the *most probable* distance r can readily be shown to be equal to

$$r = (n)^{1/2}l$$

where l is the length of each step. The random walk is used to estimate the end-to-end distance of a polymer chain; it also describes the path of a particle under **Brownian motion**. The random walk may be extended to three dimensions; see **random coil**.

Rayleigh-Gans-Debye scattering   An improved version of the **Rayleigh scattering** theory, valid in the region $d < \lambda$. In this theory the forward and backward scattering are not in general of the same magnitude, so that the **dissymmetry ratio** can be used to estimate particle size.

Rayleigh, Lord (John William Strutt) (1842-1919).   English physicist. Awarded Nobel Prize 1904.

Rayleigh ratio   A quantity obtained from **light-scattering** data, which is proportional to the spectrophotometrically-deter-mined**turbidity**.

Rayleigh scattering   Light scattered from particle arising from the interaction of the electromagnetic field of the radiation with the particle. Rayleigh scattering from the molecules of the atmosphere is responsible for the blue color of the sky. The Rayleigh equation for this scattering is valid only in the approximation $d \ll \lambda$, where d is the diameter of the scattering particle, and λ is the wavelength of the incident light. See light scattering; Rayleigh-Gans-Debye scattering.

Rayleigh-Taylor instability   The condition under which an interface is unstable owing to acceleration of the interface perpendicular to its plane, and directed from the lighter into the denser phase. Disruption of the interface will occur under the condition

$$(2\pi/\lambda)^2 < (\rho_1 - \rho_2)a/3\gamma$$

where λ is the wavelength of the disturbance, ρ_1 and ρ_2 are the densities of the two phases, γ is the interfacial tension, and a is the acceleration. See **capillary ripples**.

receding contact angle   The **contact angle** observed as the meniscus contracts as a result of withdrawing liquid from the surface; also the smaller of the contact angles observed if the solid surface is tilted.   See **advancing contact angle; contact angle hysteresis**.

reduced viscosity   Viscosity of a solution, defined as

$$\eta_{red} = (\eta/\eta_0) - 1/c$$

where η is the solution viscosity, η_0 is the solvent viscosity, and c is the concentration.   IUPAC name - *viscosity number*.

regioselective *adj*   Of a **catalyst** by means of which reaction takes place at a specific molecular site.

relative viscosity   Ratio of the viscosity of a solution to that of the solvent

$$\eta_r = \eta/\eta_0$$

IUPAC name - *viscosity ratio*.

relaxation effect In **DLVO theory**, the time required for the restoration of the charge distribution symmetry around a colloidal particle whose centers of positive and negative charge have been disturbed by, e.g., particle motion.

relaxation time The time that it takes for some exponentially decaying property, e.g., **viscoelastic** deformation, to decrease to $1/e$ (e = base of the natural logarithms) of its original value, i.e., by approximately 36.8 per cent.

repeptization *n* The reversal of the **coagulation** of a sol by dilution of the coagulating electrolyte. —**repeptize** *v*. See **peptize.**

resolution *n* See **resolving power.**

resolving power The ability of a microscope to reveal closely adjacent structural details as actually separate and distinct. For an optical microscope, the resolving power is slightly greater than the wavelength of the incident light divided by twice the **numerical aperture.** Similar relations exist for other types of microscope, e.g., *electron microscope.*

retardation *n* The reduction in **interaction energy** which occurs at particle separations of more than about 50 nm, arising from the fact that the electric field can only be propagated from one molecule to another at the speed of light. —*adj* as in *retardation effect.*

required HLB The **HLB** of an **emulsifier** required for the emulsification of a particular liquid phase. See **hydrophile-lipophile balance.**

reverse osmosis See **ultrafiltration.**

rheology *n* The study of the deformation and flow of matter; the measurement of flow properties. —**rheological** *adj.* See **viscosity.**

rheometer *n* A device for the measurement of **rheological** properties (also formerly used for a device for the measurement of electric current).

rheopexy *n* The property of hardening or becoming more viscous on gentle stirring or shaking. —**rheopectic** *adj.* See **dilatancy, rheology, thixotropy.**

Richardson equation For the viscosity of an emulsion

$$\eta = \eta_0 e^{k\phi}$$

where η and η_0 are the viscosity of the emulsion and of the **continuous phase,** respectively, and ϕ is the **volume fraction.**

root mean square The square root of the arithmetic mean of the squares of a series of numbers. *abbr* rms. See Appendix B.

root mean square distance The end-to-end distance given by a **random walk,** which is the **root mean square** of the individual steps. See **random coil.**

S

salting out, salting in   The effect of added electrolyte on the properties of a **nonionic surfactant**, arising from the competition of the surfactant (principally the head group) and the electrolyte for association with the water.   In *salting in* the solubility of the surfactant in effect increases, and the **cmc** increases, vice versa for *salting out*.

scanning electron microscope; scanning electron microscopy See SEM.

scanning tunneling microscopy   See STM.

scattering efficiency   See scattering factor.

scattering factor   A dimensionless quantity Q_{sca} by which the

139

geometric cross-section of a particle is multiplied to obtain the effective blocking power to light. It is a function of the ratio R/λ where R is the geometric radius of the particle and κ is the wavelength of the light.

scattering, light See **light scattering; PCS; Rayleigh-Gans-Debye scattering; Rayleigh scattering.**

Schiller layers Layering of settled colloidal particles, e.g., Fe_2O_3 and WO_3 sols, characterized by iridescence. The formation of such layers is taken as evidence of long-range attractive forces.

schlieren optics A device for measuring the gradient of refractive index, and thus the concentration, in a settling or centrifuged system. It is extremely sensitive to the presence of a boundary between a slowly and more rapidly settling material. From Ger. plural *Schlieren* of *Schliere* streak. See **ultracentrifuge.**

Schottky defect A crystal lattice **defect** originating from lattice vacancies.

Schulze-Hardy rule The observation that the stability of a colloidal system is proportional to the valency of the **counterions.** Simple **DLVO theory** shows that the **critical coagulation concentration** is proportional to the inverse sixth power of the valence. However, this applies to particles with a high **surface potential**; for low surface potentials the **CCC** is proportional to the fourth power of the surface potential and inversely proportional to the square of the counterion valency. Named for H.

Schulze and W. B. Hardy (1864-1934).

schutzkolloid *n*   From Ger. *Schutz-* protect + *Kolloid*. See **protective colloid**.

secondary ion mass spectrometry   See **SIMS**.

secondary minimum   In **DLVO theory**, a potential minimum at a finite distance of separation, corresponding to a metastable state, e.g., **flocculation**. See **primary minimum; pseudosecondary minimum**.

second virial coefficient   See **virial coefficient**.

sediment *n*   A layer of particles or other matter that settles to the bottom of a liquid. —**sediment** *v*.

sedimentation *n*   The settling of dispersed particles under the influence of gravity. The *sedimentation rate* is governed by Stokes' law.

sedimentation coefficient   The quantity used in measuring sedimentation in a centrifugal field (e.g., in **ultracentrifugation**)

$$s = (dR/dt)/w^2R$$

where dR/dt is the sedimentation rate and w is the angular velocity in the centrifuge, i.e., the ratio of the sedimentation velocity to the centrifugal acceleration. See **svedberg**.

sediment volume The volume occupied by a sediment, which is controlled by the probability of adhesion between the particles.

seed *v* To induce **nucleation** by the addition of small crystals of the desired product or other material of the same crystal habit.

self-assembly *n* See **aggregation; micellization.**

self-diffusion *n* **Diffusion** of a particle in a concentrated isotropic dispersion of similar particles. See **diffusion coefficient; Einstein equation, for diffusion.**

SEM *acronym* *s*canning *e*lectron *m*icroscope. An instrument in which a beam of electrons from a tungsten filament is focussed by a system of magnetic lenses on an area of surface of 5-15 nm in diameter in an evacuated chamber. The secondary electrons emitted by the sample are collected at a positive electrode, and the resulting signal is amplified to modulate the display on a cathode ray tube. By *scanning* the surface with the electron beam a high-**resolution** image of the surface is displayed.

semipermeable membrane A **membrane** or film permeable only to small molecules; thus, capable of sorting small molecules or particles by size, as in **osmosis.**

sensitize *v* To make a system more susceptible to electrolyte by adsorbing a polyelectrolyte. See **peptize.**

septum *n* In foams, the film which separates individual bubbles. The point at which three septa meet is the **Plateau bor-**

der. —septa *pl*.

sessile drop   A drop of liquid in its equilibrium position on a surface. From L. *sessilis* fit for sitting upon.

sessile drop method   A method for the determination of **surface tension** in which the shape of a drop or a bubble resting on a plane is analyzed in terms of its equatorial diameter and the height of the equatorial diameter above the plane. Application of the tables of **Bashforth** and **Adams** permits calculation of the surface tension. The term is also applied to the measurement of **contact angle** in which the angle between the drop and the surface is directly measured.

settling radius   The radius of a particle (assumed to be a sphere) determined by measurement of the rate of settling under gravity. The method is applicable to particles which are sufficiently large that **Brownian motion** is negligible. See **equivalent settling radius; sedimentation.**

shear   *n*   The lateral deformation produced in a body by the application of an external force. —**shear** *v*.

shear modulus   The ratio of the **shear stress** (the applied force per unit area) and the **shear** (or *shear strain*) produced by the force. See **rheology.**

shear rate   The time rate of application of **shear** force.

Sibree equation   For the **viscosity** of an emulsion

$$\eta = \eta_0[1/1 - (h\phi)^{1/3}]$$

where η and η_0 are the viscosities of the emulsion and of the **continuous phase**, respectively, ϕ is the volume fraction, and z is the so-called *volume factor*. See **Hatschek equation**.

simple theory　An oxymoron.

SIMS *acronym* *s*econdary *i*on *m*ass *s*pectrometry. If a surface is irradiated by an electron beam with an energy greater than about 3 kV, secondary ions are emitted, which may be analyzed by mass spectrometry.

Singer equation　Relates the **surface pressure** of a polymer, e.g., protein, **monolayer** to the structural properties of the polymer

$$\pi = -(kT/b)\{\ln(1-b/\sigma) - [(n-1)/n](z/2)\ln(1-2b/z\theta\}$$

where n is the number of links in the polymer chain (amino acid residues for a protein), z is a measure of the chain flexibility (where $z=2$ represents a rigid chain and may be as high as 4 for a flexible one), b is the close-packed area of a chain segment, and σ is the average area per segment. Named for S. J. Singer.

sinter **v**　To agglomerate or cement small particles of metal or ceramic under the influence of heat (and pressure). The effect is due to fusion and inter-diffusion at localized centers where the radius of curvature is very small, and implies some degree of mobility in the surface atoms. From Ger. *sinter* dross; related to *cinder*.

SI (Système International) Units The system of units advocated by IUPAC; some important constants are given in SI units in Appendix A.

six-twelve law See **Lennard-Jones potential.**

size exclusion chromatography See **gel permeation chromatography.**

skewness A measure of the "lopsidedness" of a distribution, given by the third moment of the distribution. See Appendix D.

slow flocculation Flocculation process in which there is a **potential barrier** opposing flocculation. See **Smoluchowski equation.**

smoke *n* A solid-in-gas dispersion; an **aerosol.**

Smoluchowski, Marion von (1872-1917) German scientist.

Smoluchowski equation There are, in fact, a number equations which are referred to under this name. However, they all in effect derive from Smoluchowski's relation for the rate of diffusional encounters R of spherical particles

$$R = 16\pi Drn^2$$

where n is the number of particles per cubic centimeter, and r and D are their radius and **diffusion coefficient**, respectively. By substituting the **Stokes-Einstein** relation for the diffusion coefficient of spheres

$$D = kT/6\pi\eta r$$

where η is the viscosity, the Smoluchowski equation becomes

$$R = 8kTn^2/3\eta$$

and the Smoluchowski equation for the rate of **coagulation** becomes

$$dn/dt = -(8kT/3\eta)n^2$$

in the absence of any **potential barrier**. See **electrophoresis; zeta potential**.

soap *n* The alkali or alkali metal salt of a fatty acid; sometimes used generically to mean any **surface-active agent**. This latter usage should be discouraged.

soil *n* In **detergency**, the "dirt" which is to be removed by washing.

soil, standard In studies of **detergency**, natural soil is simulated by a standard, synthetic soil, to overcome the problems associated with the complexity and non-reproducibility of natural soil.

sol *n* Any fluid **colloidal** system (usually transparent); e.g., a protein sol, a gold sol, etc.

solubility parameter A measure of solubility equal to the square root of the **cohesive energy density**. If two substances have solubility parameters which are equal (or at least very close), they will be mutually soluble. The greater the difference, the less the mutual solubility. Due to **J. H. Hildebrand.**

solubilizate *n*   A substance subject to **solubilization.**

solubilization *n*   The bringing into solution of insoluble organic substances by **surfactant** solutions above the **cmc.** The process involves incorporation of the **solubilizate** into the surfactant **micelle.** —**solubilize** *v.*

solution *n*   A molecular dispersion.

sorption *n*   A technical term used to describe a still unknown mechanism of adsorption (coined by **J. W. McBain**).

sorptive *n*   The surface-active molecules contained in the bulk phase (rarely used).

specific surface area   The area of a particle or **floc** per unit of weight.

specific viscosity   A representation of the viscosity of a solution

$$\eta_{sp} = \eta/\eta_0 - 1$$

where η is the viscosity of the solution and λ_0 is the viscosity of the solvent.

spheroid *n*   A solid geometrical figure similar in shape to a sphere, as an **ellipsoid.** See **oblate; prolate.**

spinning drop tensiometer A device for determining very low **interfacial tensions** by introducing a drop of one liquid into a tube (parallel to the ground) containing the other liquid, and rapidly rotating the tube. The droplet takes on an ellipsoidal shape, and the long axis r_0 of the drop is measured. The interfacial tension is given by

$$\gamma = \omega^2 \Delta \rho r_0^{3/4}$$

where ω is the speed of rotation, and $\Delta \rho$ is the difference in density between the two liquids. This relation is called the **Vonnegut** equation. The method is apparently capable of measuring interfacial tensions as low as 10^{-6} mN/m^2, and possibly lower.

spinode *n* A point (for example, in a **phase diagram**) where two branches of a curve meet, end, and are tangent; a cusp.

spinodal *adj* *Spinodal* is sometimes used as a noun, probably as an ellipsis for *spinodal point*. Such usage is to be discouraged.

spread *v* The behavior of a liquid on a solid or another liquid when the contact angle is zero.

spreading coefficient A measure of the ability of a liquid to spread on the surface of another liquid or a solid. It is defined as the difference between the work of adhesion between the two phases and the work of cohesion of the spreading liquid

$$S = W_A - W_C$$

which can be written in the form

$$S = \gamma_1 - \gamma_2 - \gamma_{12}$$

ψθϵσϵ γ_1 is the **surface tension** of the solid or liquid substrate, γ_2 is the surface tension of the spreading liquid, and γ_{12} is the **interfacial tension**. If $S > 0$, spontaneous spreading occurs. Generally ascribed to W. D. **Harkins**, but first stated (in words) by C. D. **Marangoni**.

spreading pressure   The lowering of the surface tension of a surface by the spreading of a monolayer, as on a **Langmuir trough**. See **surface pressure**.

sputtering *n*   A process which employs ions of an inert gas to dislodge atoms from the surface of a crystalline solid (e.g., a metal such as gold), which are then electrically deposited to form an extremely thin film on the surface of a solid; used in the preparation of samples for various types of surface examination, especially TEM and SEM.

stability *n*   The ability of a **colloidal dispersion** to maintain a uniform distribution of particles throughout the dispersion. —**stabilize** *v*.

stability ratio   A measure of the effectiveness of the potential barrier in preventing **coagulation** of particles. The ratio W is defined as the ratio between the number of collisions between particles to the number of collisions resulting in coagulation.

stabilization, steric   See **steric stabilization**.

stalagmometer *n*   A device for determining **surface tension** by measuring the number (or weight) of drops in a given quantity of liquid. From Gr. *stalagmos* dripping. —**stalagmometry** *n*. —**stalagmometric** *adj*.

standard deviation   See Appendix B.

standard distribution   Normal distribution. See Appendix D.

static mixer   Device for the preparation of fine **dispersions** or **emulsions**, in which the substances to be mixed are pumped at a fairly high velocity through a tube, the interior of which is fitted with regularly-spaced flights. So-called because it has no moving parts.

Staudinger, Hermann (1881-1965).   German chemist. Pioneer in the study of the physical chemistry of polymers. Nobel Prize 1953.

Staudinger-Mark-Houwink equation   The relation between the intrinsic viscosity of a polymer solution and the molecular weight of the polymer

$$[\eta] = KM^a$$

where M is the molecular weight, and K and a are constants characteristic of the polymer, and which are determined experimentally.

step *n*   See **defect**.

steric stabilization The stabilization of a colloidal particle by the adsorption of polymer to the surface; the stabilization arises from the interaction of the polymer chain with the **continuous phase** and by the interaction of the polymer chains on adjacent particles. Strictly speaking, the effect is not *steric*, but is thermodynamic in origin.

stereoselective *adj* Of a catalyst by means of which reaction is induced in a particular molecular orientation.

Stern isotherm An extension of the **Langmuir isotherm.**

Stern layer In **DLVO theory**, the region near the interface in which adsorbed ions form an inner compact layer, beyond which is the diffuse **Gouy layer.** Named for **O. Stern.**

Stern, Otto (1888-1969) German-American physicist. Nobel prize 1943.

sticking coefficient The fraction of gas molecules arriving at a surface per unit time which adhere to the surface; the probability that a molecule will **adsorb.**

Stirling approximation For the approximate calculation of the logarithm of factorials

$$\ln n! = n \ln n - n$$

useful in calculations where n is large. After James Stirling (1692-1770) Scottish mathematician.

STM *acronym* *s*canning *t*unneling *m*icroscopy. A technique for the examination of surfaces which employs a needlelike probe that scans the surface at a distance of about 100 nm. A current passes between the probe and the surface when a voltage is applied to the probe, and the size of the current, which varies with the gap between the two surfaces is an indicator of the distance between them.

Stokes, Sir George Gabriel (1819-1903)   British mathematician and physicist.

Stokes' law   **Sedimentation rate** under the influence of gravity is governed by Stokes' Law

$$u = 2gr^2(d_1 - d_2)/9\eta_2$$

where u is the velocity of sedimentation, r is the radius of the sedimenting particle, d_1 is the density of the particle, d_2 and η_2 are, respectively, the density and viscosity of the suspending liquid, and g is the acceleration of gravity. If the density difference is negative, upward sedimentation, or **creaming**, takes place. Strictly, the equation holds only for **monodisperse** rigid spheres. See **Rybczynski-Hadamard equation.**

Stokes-Navier equation   See **Navier-Stokes equation.**

Stokes settling radius   See **settling radius.**

strain *n*   The deformation response to an applied **shear.**

strain rate   See **shear rate.**

streaming current In electrophoresis, that part of the current due to the net displacement of the mobile part of the **double layer** relative to the stationary part.

streaming potential The **zeta potential** obtained from measurements of **electro-osmosis**.

streamline flow The flow of a fluid (for example, past an object) such that the velocity of fluid at any fixed point is constant or varies in a uniform manner.

stress *n* A force applied to a body to bring about deformation. Stress may be compressive, tensile, or **shear**. —**stress** *v*.

substrate *n* The adsorbing surface, which may be liquid or solid (note that in medical and biochemical terminology *substrate* may have a quite different meaning).

suction pressure The *negative* **Laplace pressure** across a liquid meniscus in a capillary. Considering this pressure as negative is simply a heuristic device, and has no special significance.

superconductivity *n* The condition under which an electrical current flows against little or no ohmic resistance. It was formerly believed that this phenomenon could occur only at temperatures equal or near to absolute zero; it has been found that certain crystalline substances exhibit superconductivity at higher temperatures (which, however, are still much below room temperature).

superconductor *n* A substance which can exhibit **superconduc-**

tivity.

surface; interface *n* A boundary between two phases. The terms *surface* and *interface* are often used interchangeably, but interface is preferred for the boundary between condensed phases, i.e., liquid/liquid, liquid/solid, etc. Thus, the term surface is reserved for the situation where one phase is a gas. The meanings of the abbreviations S/L, S/G, L/L, L/G, and S/L/G are obvious; surfaces or interfaces may also be abbreviated in terms of the chemical nature of the phases, e.g., W/O for water/oil.

surface-active agent A chemical compound characterized by the presence of **hydrophilic** and **lipophilic** moieties. As a consequence, the surface-active molecules readily adsorb at an **interface**. See **adsorption; detergent; emulsifier; surfactant.**

surface activity Broadly, any phenomenon occurring specifically at a **surface**; specifically, the behavior of **surface-active agents.**

surface analysis Broadly, any technique for the determination of the chemical or physical composition of a surface. See **AES; SEM; LEED.**

surface average diameter See Appendix B.

surface charge density The quantity of charge per unit area of surface.

surface concentration See **surface excess.**

surface excess   The **surface concentration** in an adsorbed **monolayer**, as defined by **J. W. Gibbs**. See **Gibbs adsorption equation**.

surface excess isotherm   For **adsorption** from binary liquid mixtures is given by

$$n_0 \Delta x_1 / m_a = n_1^s x_2 - n_1^s$$

where n_0 are the total number of moles *before* adsorption, x_1 and x_2 are the mole fractions of components 1 and 2 in solution at equilibrium, n_1^s and n_2^s are the number of moles of components 1 and 2 on the surface of unit mass at equilibrium, and $\Delta x_1 = x_{0,1} - x_1$, where $x_{0,1}$ is the mole fraction of component 1 in solution before adsorption. Also **composite isotherm**.

surface mean diameter   See Appendix B.

surface of tension   A mathematical construct which replaces the rather *fuzzy* real surface of, say, a droplet by a sharp boundary. It is at this surface that the **surface tension** is assumed to exist.

surface pressure   The two-dimensional pressure exerted by a **monolayer** against a barrier, as in a **film balance**, defined by

$$\pi = \gamma_0 - \gamma$$

where γ_0 is the surface tension of the clean surface and γ is the surface tension of the monolayer-covered surface.

surfactant *n*   A neologism, coined (possibly) by F. D. Snell

(1898-1980), to mean **surface-active agent.**

surface viscosity The two-dimensional analogue of viscous flow exhibited by a monolayer, either adsorbed or deposited. There are two types of viscosity, *dilational* and *shear*. In the case of monolayers, the surface dilational viscosity κ is defined by

$$\Delta\gamma = \kappa(1/A)\mathrm{d}A/\mathrm{d}t$$

where ψ is the **surface tension**, A is the area, and t the time. Thus, $1/\Theta$ is fractional change in area per unit time per unit applied surface pressure. The surface shear viscosity λ_s is defined by the relation

$$f = \lambda_s l \; \mathrm{d}v/\mathrm{d}x$$

where f is the force required to move two line elements of length l relative to one another with a velocity gradient $\mathrm{d}v/\mathrm{d}x$.

surface viscometer Device for measuring either dilational or shear **surface viscosity.** Devices for measuring dilational surface viscosity depend on the measurement of **surface tension** as the monolayer is compressed at a uniform rate; shear surface viscosity may be measured by devices analogous to the three-dimensional **couette** or **capillary viscometer.**

suspendant *n* An agent used to stabilize a suspension or dispersion (rare).

suspension *n* A finely divided solid suspended in a liquid medium. See **dispersion.**

svedberg A unit of sedimentation equal 10^{-13} time the sedimentation coefficient. Named for The Svedberg.

Svedberg, The (1884-1971) Swedish chemist, noted for development of the ultracentrifuge. Nobel Prize 1926.

swamping electrolyte A large excess of electrolyte, which *swamps* out effects due to electrolytic charge, e.g., in the determination of molecular weight by measurement of osmotic pressure.

swelling *n* Uptake of liquid or a gas by a gel or solid without increase of volume.

swelling pressure The pressure difference between a gel and its equilibrium liquid, which prevents further swelling.

swollen micelle See micelle, swollen.

syneresis *n* Spontaneous shrinking of a gel with exudation of liquid.

Szyszkowski equation A semi-empirical equation for the concentration dependence of the surface tension, usually for aqueous solutions

$$\gamma/\gamma_0 = 1 - b \ln(1 + C/a)$$

where γ is the surface tension of the solution, γ_0 is the surface tension of the solvent, b is a constant characteristic of a homologous series of organic compounds, a is a constant characteristic of each compound, and C is its concentration. Named for Bogdan von Szyszkowski (1873-1931).

T

tactoid *n* A roughly (American) football-shaped **aggregate** found in solutions of macromolecules, e.g., proteins. Two kinds of tactoids are found; in a solvent-rich part of the system, *positive* tactoids are made up of the solute, while in the solute-poor region, *negative* tactoids of solvent may exist.

Tafel equation The so-called first law of **electrode** kinetics, relating **overvoltage** η to the current density i

$$\eta = a - b \ln i$$

where a and b are empirical constants. Named for J. Tafel.

tails, polymer For an adsorbed polymer, the chain ends extending into the solution. See **loops, polymer; trains, polymer.**

Tamman temperature   The temperature ($\approx 0.5T_M$, where T_M is the melting point) at which the rate of **sintering** becomes appreciable.

Tate's law   The weight W of a freely-falling drop formed at the end of a tube is given by

$$W = 2\pi r\gamma$$

where r is the radius of the tube, and γ is the **surface tension**. This is the basis for the measurement of surface tension by the **drop weight method**.   Named for Thomas Tate.

Taylor's equation   For the **viscosity** of an **emulsion**, taking into account the viscosity of the **internal phase**

$$\eta = \eta_0[1 + 2.5\phi(2.5p + 1/p + 1)]$$

where $p = \eta_i/\eta_0$, and where η, η_i, and η_0 are the viscosity of the emulsion, the viscosity of the internal phase, and the viscosity of the **continuous phase**, respectively.   Thus, the coefficient of the **volume fraction** varies from 1 to 2.5 between $p = 0$ and $p = \infty$.

tears of (strong) wine   The phenomenon of droplets of liquid flowing down the walls of a partially-filled glass of strong wine, e.g., port or brandy, arising from the differential evaporation of alcohol and water, especially in the region of the **meniscus**.   Since the alcohol (with a lower surface tension) evaporates more rapidly, the **surface tension** of the liquid in the meniscus increases, and the meniscus crawls up the wall, dragging the strong wine with it, until a droplet separates and runs down the wall.* See **Marangoni effect.**

Teller, Edward (b. 1908) Hungarian-American physicist. See **BET isotherm.**

tenside *n* A surface-active agent.

tensile strength The resistance of a substance to longitudinal stress, defined as the minimum amount of longitudinal stress required to rupture the material. The hypothetical tensile strength of a liquid is often assumed to be equal to twice the surface tension. See **cohesion, energy of.**

tensiometer *n* A device for the measurement of **surface** and **interfacial** tension. See **duNoüy tensiometer; spinning drop tensiometer; stalagmometer.**

terrace *n* See **defect.**

tessellation *n* A tessellated pattern; a mosaic in which the pattern is repeated. Also —*v* The act or art of tessellating. From L. *tesselatus* little cube, ult. from L. *tessera* tile, little cube.

tetrahedral angle The angle formed by the faces of a tetrahedron, equal to 109°28′. In foams, the angle formed by the intersection of four **Plateau borders.**

thermodynamic stabilization See **steric stabilization.**

---

* I believe that I have read somewhere that this phenomenon was first described by Leonardo da Vinci, but I cannot locate the reference.

theta conditions The state of a polymer solution at the **theta temperature.**

theta solvent A solvent at the **theta temperature.**

theta temperature For a polymer solution, the temperature at which the **second virial coefficient** is zero; i.e., the temperature at which solvent effects on coil dimensions just compensates volume exclusion, so that the solution appears to exhibit ideal behavior. Also **Flory temperature.**

thixotropy *n* The property of becoming less viscous when stirred, owing to the breakdown of an internal structure, as in a gel. On cessation of stirring, the structure reforms and the viscosity increases. —**thixotropic** *adj*. See **rheology; rheopexy.**

Thomson, William (1824-1907) British scientist. First Baron **Kelvin.**

tilting plate A method for the determination of solid/liquid **contact angle,** in which a plate of the solid is immersed in the liquid, and its angle varied until a flat **meniscus** is obtained. The angle of tilt relative to the liquid surface is the desired contact angle.

Tiselius, Arne (1902-1971) Swedish biochemist. Nobel Prize 1948.

Tiselius apparatus Device for the measurement of **electrophoretic mobility** by the **moving boundary** method. Particularly useful in measurements on macromolecules, e.g., proteins.

topochemical *adj* Of reactions occurring on a surface at a sufficiently slow rate so that their progress can be followed.

topotactic *adj* Of reactions occurring on a crystalline surface, in which the product or products retain the external crystalline shape of the reactant crystal.

tortuosity *n* In **porosimetry**, a calibration term corresponding to the radius of an equivalent cylinder, thus characterizing the porous medium under investigation.

t-plot *n* A method of plotting **adsorption isotherms** in terms of the thickness t of the adsorbed layer.

trains, polymer In an adsorbed polymer, the segment of the polymer chain which lie flat at the surface. See **loops, polymer; tails, polymer.**

transfer coefficient In electrochemistry, that portion of the potential difference between the metal and solution phases which contributes to the activation energy.

Traube's rule The observation that in solutions of a homologous series of organic compounds $R(CH_2)_n X$, the concentration required to achieve a given **surface tension** was reduced by a factor of about 3 for each CH_2 group. **Langmuir** demonstrated that the Traube rule implied that the work required to bring one CH_2 group from the bulk solution to the surface was $RT\ln 3 = 640$ cal/mole. Named for Isador Traube (1860-1943), German chemist.

turbidity *n* The attenuation of transmitted light by **light scat-**

tering. Colloquially, the appearance of *cloudiness.*

turnover number   In catalysis, the number of molecules of reactant converted to product per second per catalyst atom.

Tyndall beam   Light scattered by small particles. Named for John Tyndall (1820-1893), British physicist. See **turbidity.**

Tyndall effect   See **Tyndall beam.**

Tyndall spectra   See **HOTS.**

U

ultracentrifuge *n* Instrument in which a cell containing the sample is rotated at very high speeds (typically, 10,000-40,000 rpm) in a horizontal cell. By the use of **schlieren optics**, the motion of dissolved or suspended molecules may be followed through the difference in optical density. Invented by T. **Svedberg**.

ultrafiltration A form of **osmosis**, in which pressure is applied to the osmotic cell, reversing the process, and, essentially causing the **semipermeable membrane** to act as a filter which separates solute from solvent molecules. Also **hyperfiltration**; re-verse osmosis.

ultramicroscope *n* A microscope, using divergent illumination, in which very small particles may be identified by means of scattered light.

ultrasonication *n* Preparation of a **dispersion** or **emulsion** by the use of ultrasound.

ultrasonic emulsification *n* See **ultrasonication**.

ultraviolet photoelectron spectroscopy See **UPS**.

unit mesh In crystallography, the two-dimensional equivalent of a unit cell.

Units, Système International See Appendix A.

UPS *acronym* *u*ltraviolet *p*hotoelectron *s*pectroscopy. A technique for obtaining information about the valence electron structure of surface atoms, in which photoelectrons are ejected from the surface. The kinetic energy of the ejected electrons is the difference between their binding energy and that of the ionizing source, low-energy ultraviolet radiation.

V

van der Waals equation See gas, two-dimensional.

van der Waals, J. D. (1827-1923) Dutch scientist, noted for his investigations into imperfect gases.

van der Waals forces See Lifshitz-van der Waals forces.

van't Hoff equation See osmosis.

Vervey, Hugo Rudolph (1893-1981) Dutch scientist. Author (with J. Th. G. Overbeek) of the *Theory of the Stability of Lyophobic Colloids* (1948), which lays the basis of the DLVO theory.

vesicle *n* A liquid droplet, stabilized by an adsorbed bilayer.

vibrating electrode method A method of measuring **surface potential** in which the electrode in air is attached to a piezo-crystal set into vibration by an audio frequency signal. The vibration of the electrode causes a corresponding change in the capacity across the air gap between the electrode and the surface. Sometimes referred to as the *vibrating plate method.*

virial coefficient The coefficients of the terms in a **virial equation**. The coefficients are usually considered to be a measure of system interactions.

virial equation An equation, such as the expanded **van der Waals** equation for an imperfect gas, which contains higher-order terms in the independent variable. Usually, only the term involving the **second virial coefficient** is considered; higher order terms are ignored, or lumped into a small correction. From L. *vis* force, strength.

viscoelasticity *n* The **rheological** behavior of a substance exhibiting a combination of viscous and elastic properties. —**viscoelastic** *adj.*

viscoelectric constant A measure of the increase of **viscosity** determined in the presence of an electric field, the fractional increase in viscosity being equal to the product of the viscoelastic constant and the square of the field strength.

viscosity *n* Resistance to flow. See **fluidity; rheology; relative viscosity; specific viscosity; reduced viscosity; inherent viscosity; intrinsic viscosity; viscoelasticity.**

viscosity average molecular weight See Appendix C.

viscosity number See reduced viscosity.

viscosity ratio See relative viscosity.

viscometer *n* A device for the measurement of viscosity. See rheology.

viscometer, surface See surface viscometer.

Vonnegut equation See spinning drop method.

Volta potential The potential just *outside* (practically, within about 10^{-3} cm) the surface of a phase. Named for Allesandro Volta (1745-1827), Italian physicist. See Galvani potential.

volume average diameter See Appendix B.

volume fraction Concentration in terms of the volumes of the components. Note that the sum of the volume fractions must equal unity.

volume, excluded See excluded volume.

von Helmholtz, Herman See Helmholtz, Herman von.

von Smoluchowski, Marion See Smoluchowski, Marion von.

W

Wagner Equation See Appendix E.

walk, random See **random walk**.

Washburn equation The velocity v of displacement of one liquid by another in a capillary is given by

$$v = r\gamma_{12}\cos\theta_{s12} / 4(\eta_1 l_1 + \eta_2 l_2)$$

where r is the radius of the capillary, γ_{12} is the **interfacial tension**, θ_{s12} is the three-phase **contact angle** (at the solid/liquid-1/liquid-2 interface), η_1 and η_2 are the viscosities of the two liquids, and l_1 and l_2 are the lengths of the liquid columns. Liquid 1 is considered to be displacing liquid 2. Named for Edward Wight Washburn (1881-1934).

water, anomalous *n*　A supposed form of water, formed spontaneously in fine capillaries; also called **polywater**, and *water II*. Although there appeared to be evidence of anomalous behavior of liquids in capillaries as early as 1955, a series of papers in the period 1962-1973, mostly from the laboratory of **B. V. Derjaguin**, presented evidence for the existence of this form of water. However, additional evidence finally appeared which demonstrated that the effects were due to material leached from the glass or silica of the capillary walls, and Derjaguin (in a classic example of the self-correcting function of science) withdrew his claims.

water flooding　See **EOR**.

water number　A titration method for characterizing **nonionic surfactants**, in which the surfactant is dissolved in 4/96 benzene/dioxane and titrated with water a persistent turbidity. It is directly proportional to **HLB**.

water repellency　The property of solid (including woven materials) which prevents penetration by a liquid, owing to the existence of a **contact angle** greater than 90°.

weight average molecular weight　See Appendix C.

weight harmonic mean diameter　See Appendix B.

weight mean diameter　See Appendix B.

wet *v* A liquid is said to wet a solid when the **contact angle** between the liquid and solid is zero. If the contact angle is greater than zero, but less than 90°, the liquid may be described as *partially wetting* the solid.

Wiener equation See Appendix E.

Wilhelmy method *n* A method of measuring **surface** (and less conveniently) **interfacial tension**, in which a **Wilhelmy plate** is suspended from the arm of, for example, a torsion balance, and then slowly brought into contact with liquid surface (or interface). The downward pull on the plate can then be shown to be equal to

$$\Delta W = \gamma p \cos \theta$$

where ΔW is the increase in weight registered by the balance, γ is the surface tension, p is the perimetric length of the plate (i.e., 2 × [length + width]), and θ is the **contact angle** of the liquid. In practice, the contact angle is usually assumed to be zero, and the width (thickness) of the plate is ignored as compared to the length. In very precise work, a buoyancy correction is required. Named for Ludwig Ferdinand Wilhelmy (1812-1864), German chemist.

Wilhelmy plate *n* A plate, usually rectangular, which may be made of sand-blasted platinum, glass, or even paper, used in the measurement of **surface** or **interfacial tension**. See **Wilhelmy method**.

Wood notation Method of labeling **LEED** patterns.

work of adhesion See **adhesion, energy of.**

work of cohesion See **cohesion, energy of.**

work function The work necessary to remove an electron from the highest populated level in a metal to a point outside. Also called *thermionic work function.*

Wulff construction A quasi-geometrical solution to the problem of defining the equilibrium shape of a crystal, i.e., the shape possessing minimum **surface free energy.**

X

x-ray diffraction   Scattering of x-rays by a crystal lattice, used to determine crystal structure by means of the **Bragg equation**.

XRD *acronym* *x-ray* diffraction.

Y

---

yield value The critical stress below which flow of a substance is not observed. Below this stress the material behaves like an **elastic solid**.

Young equation The work of **adhesion** of a liquid at a liquid/solid/vapor interface is given by

$$W_{slv} = \gamma_{lv}(1 + \cos \theta)$$

where γ_{lv} is the **surface tension** of the liquid, and θ is its contact angle with the solid. Sometimes called the *Young-Dupré* equation. Named for **Thomas Young.**

Young-Laplace equation The difference in pressure between

177

the two sides of a curved surface, e.g., a bubble, is given by

$$\Delta P = \gamma(1/R + 1/R_2)$$

where γ is the **surface tension** and R_1 and R_2 are the principal **radii of curvature.** Named for T. **Young** and P. S. **Laplace.**

Young, Thomas (1773–1829) British scientist.

Young's modulus See **Hooke's law.**

Z

zeolites *n*   Aluminosilicate mineral characterized by the presence of large cavities, arising from the way in which the (Al, Si)O$_4$ tetrahedra are linked; as a result, the windows into the cavities are also of a defined size. Consequently, zeolites may act as *molecular sieves*, and may be used for separations according to molecular size. Zeolites are also employed as **catalysts**.

zero point of charge   The pH or salt concentration at which a colloidal particle has a zero **zeta potential**, and thus does not undergo **electrophoresis**.

zeta potential *n*   The potential drop existing across the mobile part of a **dounle layer**, responsible for **electrokinetic** behavior, hence also **electrokinetic potential.** See **DLVO** theory; **Smoluchowski equation.**

Zimm plot A method of plotting **light-scattering** data for high-molecular weight substances as a function of concentration and scattering angle which permits the determination of both the molecular weight of the scattering particle and its **radius of gyration.**

Zisman plot A plot of the cosine of the solid/liquid **contact angle** of a series of liquids versus their **surface tension.** The surface tension at which the plot crosses the line for $\cos \theta = 1.0$ ($\theta = 0$), is called the **critical surface tension** of the solid.

Zisman, William Albert (1905-1986) American chemist.

zone electrophoresis Method of separation of colloidal particles or macromolecules in which the electrophoretic support medium is moist filter paper or a polyacrylamide gel. Related to *solid-liquid chromatography.*

zpc *acronym* See **zero point of charge.**

zwitterion *n* An ion containing both a negative and positive charge. Fr. Ger. *zwei,* two, through Ger. *Zwitter,* hybrid + *ion.* —*adj.* zwitterionic.

APPENDIXES

Appendix A

Useful Constants in Colloid and Surface Science[*]

Constant	Symbol	Value	Units
Acceleration due due to gravity	g	9.80621	$m\ sec^{-2}$
Avogadro number	N_A	6.02252×10^{23}	
Boltzmann constant	k	1.38054×10^{-23}	$J\ K^{-1}$
Elementary charge	e	1.69210×10^{-19}	C
Faraday constant	F	9.6485×10^4	C
Gas constant	R	8.3143	$J\ K^{-1}\ mol^{-1}$
Permittivity of free space	ϵ_o	8.854×10^{-12}	$F\ m^{-1}$
Planck constant	h	6.625×10^{-34}	$J\ s$
Velocity of light in vacuum	c	2.997925×10^7	$m\ s^{-1}$

[*] In SI Units.

183

Appendix B

Averages and Means

In colloid and surface science (as well as in polymer science) the situation often arises where, owing to non-monodispersity, a quantity cannot be represented by a single value, e.g., in a particle-size distribution. In this case, it is useful to be able to describe the quantity by an *average* or *mean* value. In addition, some measure of the spread of the values around the average is required, i.e., the *standard deviation* (See Appendix D). In ordinary discourse, when the word *average* is used, the arithmetic mean is meant. However, there are a number of other formulations of the mean which have specific applications. In fact, certain types of measurements result in a mean specific to that measurement. The averages which have been are used in colloid and surface science are listed below.

I. Generalized Mean

Except for the special cases of median, mode, and harmonic mean (described below) all means can be described by a generalized mean

$$(x)^{q-p} = \int_{x_0}^{x_m} x^q (dn/dx)dx / \int_{x_0}^{x_m} x^p (dn/dx)dx$$

where the n_i are the number in the ith interval, and the sum of all the n_i is equal to the total number of particles n. The quantities p and q are integer indices which may have zero or positive (in some cases, negative) values. The sum $p + q$ is called the *order* of the mean. When $p = 0$ and $q = 1$, the relation yields the arithmetic mean. In the definitions below, the values of p and q are given, when appropriate.

For classified data (i.e., data collected in groups of finite size, as, for example, a range of diameters), the above equation takes the form

$$\overline{x}_{qp} = \sum_i n_i x_i^q / \sum_i n_i x_i^p$$

where the n_i are the number in the ith interval, $\sum_i n_i = n$, the total number of elements. When $p = 0$, $q = 1$, Eq. 2 yields the arithmetic mean.

The generalized mean is apparently not meaningful for $p = q$. However, by application of limit theory it is possible to generate the so-called *hypergeometric* means. The most important of these is for the case $p = q = 0$, which is the *geometric* mean (one may also obtain the geometric mean from the generalized mean by replacing the variable x by log x, and setting $p = 0$, $q = 1$).

The means defined for various values of p and q are not independent. For example,

$$(\overline{x}_{qp})^{q-p} = (\overline{x})^{q-c} / \overline{x}_{pc})^{p-c}$$

where c is some arbitrary integer. In particular, for $c = 0$

$$(\overline{x}_{qp})^{q-p} = (\overline{x}_{q0})^q / (\overline{x}_{p0})^p$$

and, finally,

$$\overline{x}_{qp} = \overline{x}_{pq}$$

II. Means of General Application

Median and Mode

The *median* is the value in a set such that exactly one-half of the values lie below it, and exactly one-half above. The *mode* is the most probable value in a set, e.g., the maximum in a particle-size distribution.

Arithmetic (Number-Average) Mean (p = 1, q = 0)

$$\overline{x}_a = \sum n_i x_i / \sum n_i$$

Geometric Mean

$$\overline{x}_g = [\Pi\, n_i x_i]^{1/n} \quad (n = \sum n_i)$$

Harmonic Mean (p = -1, q = 0)

$$\overline{x}_{ha} = [\sum n_i / x_i / \sum n_i]^{-1}$$

III. Means Applicable to Diameters or Radii

Number Length Mean Diameter (p = 1, q = 0)

$$\overline{d}_{10} = \sum n_i d_i / \sum n_i$$

Number Area Mean Diameter (p = 2, q = 0)

$$\overline{d}_{20} = [)\sum_i n_i d_i^2 / \sum_i n_i]^{1/2}$$

Number Volume Mean Diameter (p = 3, q = 0)

$$\overline{d}_{30} = [\sum_i n_i d_i^3 / \sum_i n_i]^{1/3}$$

Linear Mean Diameter (p = 2, q = 1)

$$\overline{d}_{21} = \sum_i n_i^2 / \sum_i n_i d_i$$

Surface Mean Diameter (p = 3, q = 2)

$$\overline{d}_{32} = \sum_i n_i d_i^3 / \sum_i n_i d_i^2$$

Weight Mean Diameter (p = 4, q = 3)

$$\overline{d}_{43} = \sum_i n_i d_i^4 / \sum_i n_i d_i^3$$

Weight Harmonic Mean Diameter (p = -3, q = 0)

$$\overline{d}_{(-3)0} = [\sum_i n_i d_i^{-3} / \sum_i n_i]^{-1/3}$$

Appendix C

Average Molecular Weights

In a polydisperse system, such as a polymeric substance, a commercial surfactant, or a system of aggregates (e.g., micelles), it is impossible to define a single molecular weight. Rather, the molecular weight is an average, the particular form of which depends on the method by which the average molecular weight was determined. There three principal molecular weight averages. These are shown below, together with the methods by which they are determined.

Number Average Molecular Weight

$$\overline{M}_n = \sum n_i M_i / \sum n_i$$

(Colligative properties, especially osmotic pressure)

Weight Average Molecular Weight

$$\overline{M}_w = \sum n_i M_i^2 / \sum n_i M_i$$

(Light scattering, sedimentation velocity)

Viscosity Average Molecular Weight

$$\overline{M}_v = [\sum n_i M_i^{1+a} / \sum n_i M_i]^{1/a}$$

(Intrinsic viscosity, where a is the constant in the Staudinger-

Mark-Houwink equation)

Note also that the ratio $\bar{M}_w / \bar{M}_n > 1$ for a polydisperse system, and that the value of the ratio is a measure of polydispersity.

Appendix D

Distribution Functions

In a polydisperse system, the distribution of the measure of interest, e.g., molecular weight, particle size, etc., may be described in two ways: as a histogram or as a continuous function. Histograms are useful for the presentation of *classified* data, i.e., where the measure of interest is described in terms of the population in a range of values. However, theoretical considerations may require the use of a continuous function, the parameters of which can be estimated by the use of the classified data. A number of useful distributions are described in the following, but it is first necessary to define the *standard deviation*, given by

$$\sigma = [\sum_i n_i(x_i - \overline{x})^2 / \sum_i n_i - 1]^{1/2}$$

where the n_i and x_i are the population and average size in the ith class, respectively, and $\overline{x}$ is the mean of all the values. The square of the standard deviation is sometimes called the *variance* of the distribution.

I. Distribution Functions

Normal (Gaussian) Distribution

$$f(x) = [1/\sqrt{(2\pi)}\sigma] \exp[-(x - \overline{x})^2/2\sigma^2]$$

The normal distribution is symmetrically distributed around the arithmetic mean. It is not usually met in naturally-occurring

191

distributions, such as emulsions, but correctly describes the distribution of random errors of measurement.

Log Normal Distribution

$$f(x) = (1/\sqrt{(2\pi)} \ln \sigma_g) \exp[-(\ln x - \ln \overline{x}_g)^2/2 \ln^2\sigma]$$

where $\overline{x}_g$ is the geometric mean, and θ_g is the standard deviation of the geometric mean. This distribution is thus distributed about the geometric mean. Many naturally-occurring distributions are fitted quite accurately by this distribution.

Zero Order Log Normal Distribution (ZOLD)

$$f(x) = [1/\sqrt{(2\pi)} \, \overline{x}_m \, \sigma_0 \exp(\sigma_0^2/2)] \times$$
$$\exp[-(\ln x - \ln \overline{x}_m 2)/2\sigma_0^2)]$$

The ZOLD is a variation on the log normal distribution, distributed around the mode $(\overline{x}_m)$ rather than about the geometric mean, and where σ_0 is equal to the logarithm of the standard deviation of the *geometric* mean. Since the mode is the most probable value of the variable, and thus an obvious value on a distribution plot, this function is intuitively satisfying. In addition, it permits exploration of the effect of changing standard deviation without changing the mode.

Rosin-Rammler Distribution

$$f(x) = nbx^{n-1} \exp(-bn^n)$$

where n and b are constants, b being a measure of the range of particle size present and n begin characteristic of the material of the dispersion. For $n = 1$, b is the reciprocal of the mode. This function was derived for solid materials prepared by crushing or shattering, but could possibly apply to emulsions prepared by jet impingement.

Poisson Distribution

For the case where the number of particles in each class is small (i.e., each size region has a low probability)

$$f(x) = e^{-n}n^{x-1}/(x - 1)!$$

where n is the total number of particles and x are the number of particles in each size class. For the use of the Poisson distribution to calculate the distribution of ethylene oxide groups in a chain see Dictionary, **Poisson distribution.**

II. Moments of a Distribution

A useful concept is that of the *moment* of a distribution about a point. The jth moment of a distribution of x_i about a point d_0 is defined as

$$M_j = \sum_i f(x_i - x_0)^j$$

It is clear that the first moment around the origin ($j = 1$, $x_0 = 0$) is the *arithmetic mean*, while the second moment is the *variance* (σ^2). The third moment measures the *skewness* of the distribution; since it is an odd function $f(x) = -f(-x)$ and thus is equal to zero for a symmetrical distribution. The fourth moment is the *kurtosis*, which is weighted towards points distant from the mean, and therefore is a measure of the length of the tail of the distribution.

Appendix E

Dielectric Constant and Conductivity of Dispersions

Numerous equations have been proposed to describe the dielectric or conductometric properties of dispersions and emulsions. A number of the most useful are collected here.

I. Conductivity

Wagner Equation

For a dispersion of spheres:

$$\kappa - \kappa_m/\kappa + 2\kappa_m = \phi(\kappa_p - \kappa_m/\kappa_p + 2\kappa_m)$$

where κ_m, κ_p, and κ are the conductivities of the continuous phase, the disperse phase, and the dispersion, respectively, and ϕ is the phase volume of the dispersion.*

Fricke Equation

For a dispersion of ellipsoids:

$$\kappa = \kappa_m - [\phi(\kappa - \kappa_p)/3(1 - \phi)]\sum \kappa_m/[\kappa_m(1 - L_i) + \kappa_p L_i]$$

$$L_i = 1, 2, 3$$

where the L_i involve elliptic integrals of the second kind, and

---

* This nomenclature will be followed throughout Appendix E.

are functions of the semiprincipal axes of the ellipsoid.

Bruggeman Equation

$$(\kappa - \kappa_p)/(\kappa_m - \kappa_p)(\kappa_m/\kappa)^{1/3} = 1 - \phi$$

When $\kappa_m \gg \kappa_p$ (as in the case of a W/O emulsion) then

$$\kappa/\kappa_m = (1 - \phi)^{3/2}$$

Hanai Equation

$$\kappa/\kappa_m = 3\epsilon(\epsilon - \epsilon_m)/(\epsilon_p + 2\epsilon)(\epsilon_p - \epsilon_m)$$

where the ϵ are the dielectric constants and the subscripts have the same significance as those for conductivity.

At high frequency, the Hanai equation becomes

$$\kappa/\kappa_m = 1/(1 - \phi)^3$$

II. Dielectric Constant

Wiener Equation

In dilute systems, e.g., $\phi \ll 1$

$$(\epsilon - \epsilon_p)/\epsilon + 2\epsilon_m = \phi[\epsilon_p - \epsilon_m/\epsilon_p + 2\epsilon_m]$$

where ϵ_p, ϵ_m, and ϵ are the dielectric constants and the subscripts have the same significance as for the conductivities.

Bruggeman Equation

For more concentrated systems

$$(\epsilon - \epsilon_p)/(\epsilon_m - \epsilon_p) = (1 - \phi)(\epsilon/\epsilon_m)^{1/3}$$

Böttcher Equation

$$(\epsilon - \epsilon_m)/3\epsilon = \phi[(\epsilon_p - \epsilon_m)/(\epsilon_p + 2\epsilon)]$$

Appendix F

Variations of the Einstein Equation for Viscosity

As pointed out in the Dictionary, the Einstein equation for the viscosity of dispersions

$$\eta = \eta_0(1 + 2.5\phi)$$

is subject to severe limitations, applying to rigid spheres and to $\phi < 0.02$. It does become exact in a limiting form, i.e.,

$$[(\eta/\eta_0 - 1)/\phi] = [\eta_{sp}/\phi] = 2.5$$
$$(\phi \rightarrow 0)$$

where η_{sp} is the *specific viscosity*.

In order to apply the Einstein equation to systems of higher, more practical, concentrations a number of variations have been proposed in which the term in ϕ is replaced by a power series, i.e.,

$$\eta = \eta_0(1 + \alpha_0\phi + \alpha_1\phi^2 + \alpha_2\phi^3 + ...)$$

where α_0, α_1, α_2, ... are constants, and α_0 usually has the value of 2.5. Examples of this type of equation are

Guth-Gold-Simha Equation

$$\eta = \eta_0(1 + 2.5\phi + 14.1\phi^2)$$

This increases the valid concentration range to $\phi < 0.06$

Eilers Equation

For emulsions of paraffin and bitumen:

$$\phi_{sp} = 2.5\phi + 4.94\phi^2 + 8.78\phi^3$$

Oliver-Ward Equation

Derived for model emulsions of rigid spheres with a distribution of sizes:

$$\eta_r = 1/(1 - k\phi) = 1 + k\phi + k^2\phi^2 + k^3\phi^3 + \ldots$$

For the systems investigated, k had values close to 2.5.

Appendix G

Bibliography

The following books have been invaluable, both as a source of definitions and of words to be defined.

A. W. Adamson, *Physical Chemistry of Surfaces*, 3rd ed., John Wiley & Sons, New York, 1976.

T. Allen, *Particle Size Measurement*, 3rd ed., Chapman and Hall, London, 1981.

R. Aveyard and D. A. Haydon, *An Introduction to the Principles of Surface Chemistry*, University Press, Cambridge, 1973.

P. Becher, *Emulsions: Theory and Practice*, 2nd ed., Krieger Publishing Co., Melbourne, FL, 1977 (Reprint of 1966 edition).

P. Becher (Ed.), *Encyclopedia of Emulsion Technology*, Vols. 1-3, Marcel Dekker, Inc., New York, 1983, 1985, 1988.

P. C. Heimenz, *Principles of Colloid and Surface Chemistry*, 2nd ed., Marcel Dekker, Inc., New York, 1986.

G. Herdan, *Small Particle Statistics*, 2nd rev. ed., Academic Press, Inc., New York, 1960.

L. Hogben, *The Vocabulary of Science*, Stein & Day, New York, 1970.

R. J. Hunter, *Foundations of Colloid Science*, Vol. I., Clarendon Press, Oxford, 1987.

M. J. Jaycock and G. D. Parfitt, *Chemistry of Interfaces*, Ellis Horwood, Ltd., Chichester and Halsted Press, John Wiley & Sons, New York, 1981.

C. A. Miller and P. Neogi, *Interfacial Phenomena: Equilibrium and Dynamic Effects*, (Surfactant Science Series, Vol. 17), Marcel Dekker, Inc., New York, 1985.

G. A. Somorjai, *Chemistry in Two Dimensions: Surfaces*, Cornell University Press, Ithaca, 1981.

R. W. Whorlow, *Rheological Techniques*, Ellis Horwood Ltd./Halstead Press, Chichester, U.K., 1980.

...and of course

S. B. Flexner (ed.), *The Random House Dictionary of the English Language*, 2nd ed., unabridged, Random House, New York, 1987.

Notes

Notes

Notes

Notes